FIR[EFLY]

GEOGRAPHY
DICTIONARY

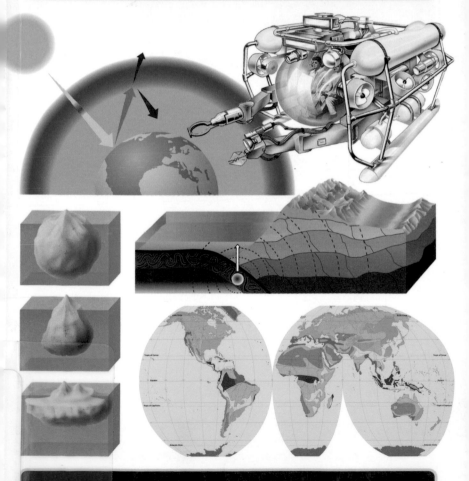

D0030672

A COMPREHENSIVE ILLUSTRATED DICTIONARY OF
PHYSICAL, HUMAN AND ENVIRONMENTAL GEOGRAPHY

DISCARD

SCC Library

DISCARD

3 3065 00357 9814

FIREFLY Santiago Canyon College
Library

GEOGRAPHY DICTIONARY

G
63
F56
2003

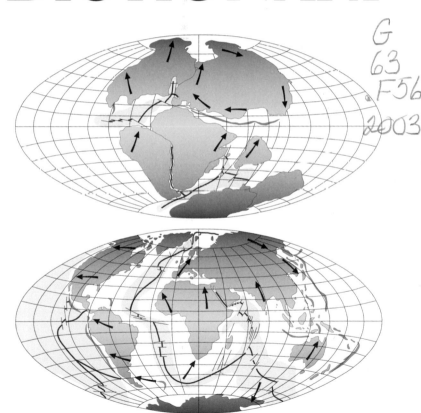

FIREFLY BOOKS

OCM53447301

Santiago Canyon College
Library

A FIREFLY BOOK

Published by Firefly Books Ltd. 2003

© Philip's 2003

All rights reserved. No part of this publication may be
reproduced, stored in a retrieval system, or transmitted
in any form or by any means, electronic, mechanical,
photocopying, recording or otherwise, without the prior
written permission of the Publisher.

First printing

National Library of Canada Cataloguing in Publication Data

    Firefly geography dictionary. – 1st North American ed.

Previous eds. published London : G. Philip under title:
Philip's geography dictionary.
Includes index.
ISBN 1-55297-838-9 (pbk.)
1. Geography--Dictionaries. I. Title: Geography dictionary.
II. Title: Philip's geography dictionary.
G63.F57 2003 910'.3 C2003-901328-6

Publisher Cataloguing-in-Publication Data
(Library of Congress standards)

Firefly geography dictionary. – 1st ed.
[240] p. ; col. ill. , maps : cm.
Originally published: Philip's geography dictionary
Summary: Geographic terms defined, plus information
tables, illustrations and maps.
ISBN: 1-55297-838-9 (pbk.)
1. Geography – Dictionaries. I. Title.
910/.3 21   G63.F52 2003

Published in Canada in 2003 by
Firefly Books Ltd.
3680 Victoria Park Avenue
Toronto, Ontario, M2H 3K1

Published in the United States in 2003 by
Firefly Books (U.S.) Inc.
P.O. Box 1338, Ellicott Station
Buffalo, New York 14205

Produced in Great Britain in 2003 by Philip's,
a division of Octopus Publishing Group Ltd,
2–4 Heron Quays, London E14 4JP

Printed in China

Using this book

The *Geography Dictionary* contains more than 1,500 articles and over 130 illustrations covering a wide range of topics in all areas of physical, human, and environmental geography. The book has been designed to be easy to use by readers at all levels.

The order of articles is strictly alphabetical. Many of the articles in this dictionary have cross-references in **bold type** to support and provide useful related information about the subject. These cross references often follow "SEE ALSO" at the end of the article.

Where appropriate, alternative names and spellings for the terms are given at the end of the article, following "ALSO CALLED". Where an alternative name is in common use, there is an entry under that name that gives a cross reference to the article that contains the definition.

The illustrations have two purposes. Firstly, they support articles, giving graphical representations and explanations of concepts. Secondly, they provide additional information, augmenting and expanding the content of the article.

A

aa The Hawaiian name for blocky lava, a type of lava flow which has solidified into a blocky mass, rather like cinders in appearance. Gases would have escaped from the molten lava in a fairly violent fashion, making the rock jagged and angular. Examples are found on Mauna Loa in the Hawaiian Islands, Mt. Etna in Sicily and Mt. Hekla in Iceland. SEE ALSO **pahoehoe**

ablation I Loss of ice in a **glacier** due to evaporation and melting. The process includes melting at the top of the ice because of sunshine, and around the edges of the ice because of rainwater or **meltwater**. **2** The loss of ice at the end of a glacier where an **iceberg** may break off.

ablation moraine The rock debris left at the side of a glacier after **ablation** has occurred.

aborigines The native inhabitants of any country, now normally used to refer to the original inhabitants of Australia only. Some Australian aborigines still live in very isolated parts of Northern Australia, but many now work on farms or live in towns. They formerly lived by hunting and gathering, using boomerangs to kill birds and animals, and were skillful at finding food in a very difficult environment.

abrasion The wearing away of rocks by the action of wind, water or ice carrying particles of dust and sand; a mechanical process, which acts rather like sandpapering. Abrasion in rivers will erode the banks and may form potholes on the bed. Abrasion is most active where there is no vegetation to protect the rocks and soil, especially in deserts. Mushroom-shaped rocks have been formed by abrasion, and telegraph poles have been worn through by wind-blown sand. Abrasion is effective along coastlines, too, where it will wear away at the cliffs as well as the shore. SEE ALSO **wave-cut platform**

abrasion platform SEE **wave-cut platform**

absentee landowner A farm owner who lives in a town and not on the farm, leaving a manager to run the estate though the owner will collect all the profits.

absolute drought A period usually of at least 15 days on none of which more than 0.01 in [0.25 mm] of rain falls.

absolute humidity The amount of water vapor present in the air at any given time. It is measured in grains per cubic foot (7,000 grains = 1 lb) or in grams per cubic meter. The amount of water vapor which the air can hold will depend on its pressure and temperature. Warm air can hold much more water vapor than cold air. For example, air at 50°F [10°C] can hold 4.1 grains per cubic foot, whereas at 68°F [20°C] it can hold 7.5 grains. When the air contains the maximum amount it can hold, it is said to be saturated and to have reached "dew point," that is, the temperature at which dew begins to be deposited. There is no specific temperature for dew point, but the relative humidity will be 100%.

abyssal Used to describe features occurring at great depths, usually over 9,840 ft [3,000 m] below sea level. An "abyssal plain" is a large, fairly level plain on the deepest parts of the ocean floor. The "abyssal zone" refers to the deepest areas of the ocean or to the area in a lake where sunlight cannot penetrate. "Abyssal deposits" are simply deposits on the deepest parts of the ocean floor.

accessibility The term used in transportation studies to indicate the ease or difficulty in getting from one place to another. Accessibility may refer to cars and roads, or to public transportation on roads or railroads. Central locations in cities are often very accessible because they have many routes. Peripheral locations may have fewer routes but may suffer less from traffic congestion. Some towns have great accessibility because of high-speed trains, and for longer distances air travel will be important.

accumulation The addition of ice and snow to a **glacier** (the opposite of **ablation**). As snow is added from snowfall and **avalanches**, it is compressed to form ice; this process usually takes place at the start of a glacier where temperatures are lower.

acid lava Lava which is viscous and slow-flowing, rather like thick oatmeal. It contains a high proportion of silica and usually solidifies quite quickly. Acidic volcanoes have much steeper cross-sections than basic volcanoes. One of the most famous of the acid lava volcanoes is Mt. Pelée in Martinique, which had a disastrous eruption in 1902. SEE ALSO **basic lava, viscous lava**

acid rain Rain which contains pollutants, in particular sulfur and nitrogen oxides. The pollutants result from the burning of **fossil fuels**, especially in thermal power stations, but they can also be caused by factories and car exhausts. Once they are in the **atmosphere**, pollutants can be washed down to Earth by rainfall as dilute forms of sulfuric and nitric acid. Although dilute, they are powerful enough to poison lakes and rivers, kill trees and speed up the erosion of buildings. Most industrial countries now have strict regulations to control the amount of pollution which is sent into the atmosphere. Winds often blow the pollution away from the countries which caused it; thus Norway and Sweden frequently suffer from Russian, German and British pollution.

acid rock Any of the **igneous** rocks which contain a high proportion of silica (more than 66%). The main silica mineral

ACCESSIBILITY

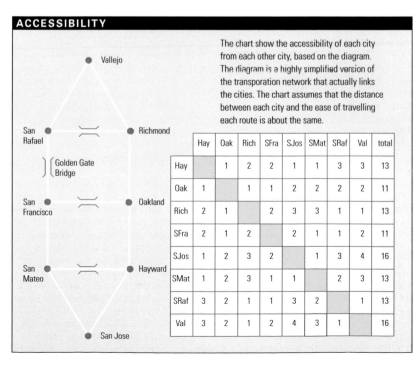

The chart show the accessibility of each city from each other city, based on the diagram. The diagram is a highly simplified version of the transporation network that actually links the cities. The chart assumes that the distance between each city and the ease of travelling each route is about the same.

	Hay	Oak	Rich	SFra	SJos	SMat	SRaf	Val	total
Hay		1	2	2	1	1	3	3	13
Oak	1		1	1	2	2	2	2	11
Rich	2	1		2	3	3	1	1	13
SFra	2	1	2		2	1	1	2	11
SJos	1	2	3	2		1	3	4	16
SMat	1	2	3	1	1		2	3	13
SRaf	3	2	1	1	3	2		1	13
Val	3	2	1	2	4	3	1		16

is **quartz**, but some **feldspars** are also quite rich in silica. Acid igneous rocks are usually light in color. The most common is **granite**. When they are weathered to form soil, they are generally infertile.

acid soil Soil which is short of bases and has a pH of 6 or less. In wet and cool areas, water leaches out all the soluble bases, especially calcium, to form acid soil. SEE ALSO **podzol**

acre An imperial unit of area covering 4,840 sq yds. It was based on a furlong [220 yds] multiplied by a chain [22 yds]: 640 acres equal 1 sq mile, and 1 acre equals 0.4047 hectares. One acre is about half the size of a football field.

acre foot (*in irrigation*) The volume of water required to cover an area of 1 acre to a depth of 1 ft, which is equivalent to 43,560 cu ft or 1,233.5 cu m.

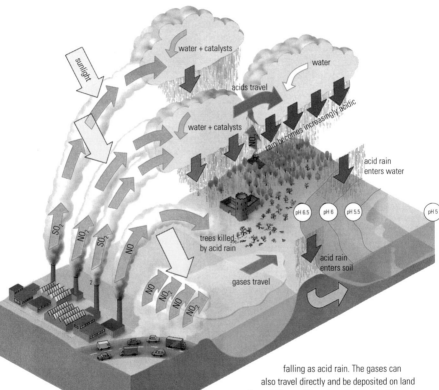

▲ **acid rain** Factory chimneys emit sulphur dioxide (SO_2) and the nitrogen oxides (NO and NO_2). Vehicle engines emit nitrogen oxides. If they are in the atmosphere for any time, the gases will combine with water to form sulphuric acid (H_2SO_4) and nitric acid (HNO_3). The acids may travel great distances before falling as acid rain. The gases can also travel directly and be deposited on land and vegetation and in water, where again they may form acids. Catalysts such as hydrogen peroxide, ozone, and ammonium promote the formation of acids in clouds. Sunlight promotes the formation of these catalysts – for example, ozone from hydrocarbons emitted by vehicle engines. As acid rain falls or drains into the lake, the acidity of the lake increases, which kills both plants and animals in the water. When acid rain falls on the ground, it changes soil chemistry which in turn can affect lakes.

active layer The layer of soil located above the **permafrost** (in **tundra** areas) which experiences periodic thawing, either daily or only in summer.

adiabatic Denoting changes in the pressure and temperature of a parcel of air, when heat from outside is not added or removed. The air will be either expanding or contracting, thus producing changes of pressure and temperature. SEE ALSO **dry adiabatic lapse rate, saturated adiabatic lapse rate**

adit A type of mining in which the miners dig in horizontally from the side of a valley. It is cheaper than having to sink a deep shaft, but can be used only when erosion has exposed the coal seam.

adobe Mud which has been dried in the sun and used as bricks. Adobe bricks are used in Mexico, Argentina and other parts of South America. Similar mud bricks, often containing straw to make them stronger, are used in many parts of Africa. Heavy rain will gradually soften them, and so adobe houses tend not to last for a long time.

adret The south-facing slope, or sunny side, of a valley, especially in the French Alps. It is the side of the valley on which most houses will be built, and also where the best farming can take place. SEE ALSO **ubac**

advection The horizontal movement of air, ocean currents or **surface runoff**.

advection fog A **fog** which forms when a warm and moist air stream moves horizontally over a cooler surface, causing its lower layer to be chilled below dew point.

aerial photography A photograph taken from the air showing the characteristics of an area, particularly useful in the making of maps. It may be taken at an oblique (i.e. slanting) angle or a vertical (i.e. straight down – a "bird's-eye view") angle.

aerobic A term denoting organisms which live in a location with free oxygen. SEE ALSO **anaerobic**

afforestation The planting of trees, generally in an area which has not had trees in the past. For example, millions of trees were planted on the dry **steppes** of Kazakhstan. Trees may be planted in order to produce wood or pulp some time in the future, or as **windbreaks**, or to hold the soil together and reduce **soil erosion**. In some **Third World** countries, trees are now being planted to provide firewood. ALSO CALLED **forestation** SEE ALSO **reforestation**

age dependency That part of a population too young or too old to be in full-time employment, who depend upon the support of those in employment. An aging population results in fewer workers supporting increasing numbers of old people.

age-sex pyramid A **frequency distribution** histogram which shows the age and sex composition of a population. The horizontal bars are usually indicated as percentages of the total population, but can be drawn as actual numbers, or as a percentage of just the male or just the female populations. Each horizontal bar can represent any number of years, but it is common to draw five-year groups (called "cohorts"). In this way, the lowest bar would represent all children aged 0–4, and the next bar would show the 5–9 cohort, and the next would be 10–14 and so on. It is normal to put the males on the left of the central axis, and the females on the right. The shape of the pyramid will vary according to the population composition of the city or country which is the subject of the histogram. ALSO CALLED **population pyramid**

agglomerate A type of rock which has been formed by volcanic activity. It consists of coarse blocks of rock or lava which have been thrown out by a volcanic eruption and then cemented together in volcanic ash.

agglomeration 1 A group of houses or settlements which have spread and grown together to merge with several neighboring settlements. **2** (*in industrial geography*) A group of industries which are close to one

AGE-SEX PYRAMID

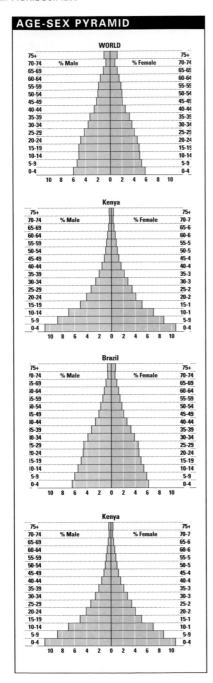

WORLD

Kenya

Brazil

Kenya

another and possibly related in what they produce or in the materials they use. An industrial agglomeration often occurs in coal-mining areas, when steel works, followed by engineering works, and then chemical works, all find that the best location for them is on the coal field, as, for example, in South Wales, northeast England, Nord in France and the Sambre-Meuse valley in Belgium. There are no industrial agglomerations in countries such as Sweden or Switzerland where hydroelectricity rather than coal was the main source of energy when industries were being developed.

agribusiness The process of running an agricultural operation as an industry – involving everything from, for example, the manufacture of the agricultural machinery to the storage and distribution of the product. SEE ALSO **commercial agriculture**

Agricultural Revolution The period in the 19th century when new machines and farming methods enabled extra food to be produced and new areas to be turned into productive land. It was not really a revolution, but merely part of the continuing change which had been affecting farming for centuries. More dramatic changes have probably occurred during the last 30 or 40 years, with many new fertilizers and pesticides, as well as new machines such as combine harvesters. SEE ALSO **green revolution**

agriculture The cultivation of the soil and the management of the natural landscape in order to grow crops, including both food and industrial crops such as cotton or rubber. There are many different types of agriculture, and the term is frequently used to include pastoral farming. SEE ALSO **commercial agriculture, extensive farming, intensive farming, subsistence farming**

Agulhas Current SEE **Mozambique Current**

A horizon The top layer of **soil** which contains fine particles of soil together with the **humus**. Much mineral and organic material, together with most of the plant roots, is found in this layer. It is the first part of any soil to be washed or blown away if **erosion** is taking place. Resting on top of the A horizon there may be an **O layer**, which is where organic material accumulates. The O layer may consist of L, which is the litter layer; F, which is the partly decomposed or fermented layer; and H, which is the well-decomposed or humified layer. SEE ALSO **B horizon, C horizon**

aid Resources (in many forms, such as money, technology, training or food and other goods) given to less-developed countries, either in times of emergencies or to promote long-term development.

air frost SEE **frost**

air mass A large body of air covering hundreds of square miles. When a northerly air mass travels over Canada and the USA, it will have come from the Arctic and will bring cold weather. Air masses are named after the region they have come from. A polar air mass will have traveled from north of the Arctic Circle, and a tropical continental air mass will have come from the Sahara, or parts of the Arabian peninsula, or Australia. All air masses gradually change as they travel, becoming wetter if they move over sea, or drier if they travel over land. If they are moving toward the Equator they will gradually get warmer, but if they are traveling toward the Arctic (or Antarctic) they will become colder.

air stream An air movement which is smaller than an **air mass**, and is simply a moving current or channel of air. The term can refer to a large wind.

alfalfa A leguminous plant, *Medicago sativa*, which is similar to clover. It has long roots and can survive spells of dry weather. It is a nitrogenous plant, which means it takes nitrogen from the air and stores it in its roots; this is then released into the soil to restore fertility. Alfalfa is a good fodder crop for cattle and is, for example, widely grown in the USA and Argentina.

algae A large and diverse group of simple plants that use **photosynthesis** to live. Algae live in aquatic habitats or in damp places on land.

algal bloom A sudden increase in numbers of aquatic **algae** resulting from changes in the chemical content or temperature of water bodies.

alkaline soil Soil with a pH of 7 or more, which usually has a high concentration of carbonates.

alluvium Fine mud or silt deposited by rivers. It accumulates on river beds but can be spread over large areas of a river valley at times of flood. All rivers have some alluvium deposits, but large rivers such as the Mississippi or Amazon have alluvial plains which are several miles wide. Alluvial deposits at the foot of steep mountains often accumulate in a triangular or cone shape, rather like an inland delta. Such deposits are called "alluvial fans" or "alluvial cones." Examples of these can be seen in many valleys in the Alps.

alp An area of grassland vegetation high up on the side of any mountain range. The alpine pasture is used by grazing animals in the summer but is snow-covered in winter. On the alp there will probably be huts or chalets, which are used by local people in the summer while looking after the cattle. In Switzerland and several other countries, some of the alps have become important locations for skiing. **The Alps** is a mountain range which extends from France through Switzerland, Italy and Germany to Austria and the former Yugoslavia.

Alpine folding The most recent large-scale example of **folding**, which occurred when the Alps, Himalayas, Rockies, Andes and other large mountain ranges were formed. SEE ALSO **fold mountains**

altiplano (*"high plateau"*) An area lying mainly in Bolivia, South America, between two prongs of the Andes Mountains. It is a closed depression, 11,500 ft to 13,000 ft

[3,500 m to 4,000 m] above sea level, filled with eroded sediments and volcanic material. It also contains many huge, flat *salars* (salt basins), which are the remnants of ancient lakes.

altitude 1 The vertical height of a point above mean sea level. **2** The vertical height of a mountain or hill measured from base to summit.

altocumulus A globular cloud formation which is found at middle altitudes (8,000 ft to 20,000 [2,400 m to 6,000 m ft]). SEE ALSO **cloud**

altostratus A gray or bluish striated, fibrous or uniform cloud layer producing light drizzle. SEE ALSO **cloud**

amphibian A group of vertebrate animals that spend the early part of their lives in water and the adult part in an aquatic or damp habitat; for example, a frog.

anabatic A term denoting a wind that blows uphill. Anabatic winds occur when the sides of a mountain valley are heated more rapidly (this usually happens in the morning) than the valley floor and the warmer, lighter air rises uphill. The opposite to anabatic is **katabatic**.

anaerobic A term denoting organisms that live without using oxygen. They will generally live in a location with little or no oxygen, usually a very wet place, such as a **peat** bog. In anaerobic conditions, dead vegetation decays very slowly and may possibly form coal or peat. SEE ALSO **aerobic**

anemometer An instrument for measuring wind speed. It consists of three cups which are blown around by the wind, and the speed of the wind can be read off the dial. More sophisticated instruments keep a written record of wind speed and can record the wind direction as well. An anemometer should be located in an open space, well away from buildings and trees, and about 33 ft [10 m] above ground level if possible.

aneroid barometer An instrument for measuring **atmospheric pressure**. Less accurate than a mercury barometer, it consists of a box with most of the air removed, the sides of which move in and out with changes of pressure. The needle recording the pressure changes can be attached to a pen which writes the pressure readings onto a piece of paper attached to a revolving cylinder. In this way, a written record of a week's pressure can be obtained. SEE ALSO **barograph**

Antarctic Circle The line of latitude at 66° 30'S. Along this line of latitude there is a 24-hour period of continuous daylight in December and a 24-hour period of continuous darkness in June: at any time, the conditions here are the reverse of those in the Arctic. Moving south toward the South Pole, the periods of continuous daylight and darkness become progressively longer. Within the Antarctic Circle lies the land mass of Antarctica. SEE ALSO **Arctic Circle**

anthracite A hard type of coal which burns with little visible flame and leaves little ash. It consists of more than 90% carbon and can be used as a smokeless fuel. Large deposits have been found in Pennsylvania in the USA and in South Wales in the UK. SEE ALSO **bituminous coal**, **brown coal**

anticline The arch of upfold in layers of rock strata. It is the result of compression, which caused the rock strata to bend. SEE ALSO **syncline**

anticyclone An area of high pressure where air is subsiding, causing dry weather. Winds are normally light and the weather is settled. Major high-pressure regions are found in the **horse latitudes**, about 20° to 30° north and south of the Equator. There are also large anticyclones in the Arctic and Antarctic. ALSO CALLED **high**

apartheid (*"separate development"*) A system of racial segregation of white and nonwhite first used in South Africa by the white government in 1948. At its most extreme, this included zoned housing within cities and segregated facilities, such as transportation. The policy was officially abandoned in South Africa after 1991, and the country's first multiracial elections were held in April 1994.

appropriate technology A term used to denote small-scale or simple tools or machines which are more useful to people in poor areas than the kind of high technology used in the developed world. For example, small hand-held plows or diggers are likely to be more appropriate for many farmers than a large tractor or combine harvester; a small loom may be more useful in an African village than a large electric-powered machine. Such machinery is easily repaired and should not be dependent on foreign experts or expensive parts. ALSO CALLED **intermediate technology**

aquifer A layer of rock which can hold large quantities of water in the pore spaces. Water may percolate along an aquifer, following the gradient of the stratum. An aquifer is generally located between two **impervious** layers.

arable A term denoting the type of farming in which crops are grown. Arable farmland is generally suitable for plowing.

Archimedes' screw A device for raising water consisting of a screw inside a tube; as the screw is turned, water passes up the thread. The lower end of the tube is placed in a river or canal, and water flows out at the upper end. It is used for irrigation in the Nile valley in Egypt and elsewhere in Africa.

archipelago A group of islands. It may be a broken chain of islands where a range of mountains has been submerged by the sea, a collection of coral islands or a group of volcanic islands, where submarine eruptions have built up islands from the ocean floor. Examples include the Aleutian Islands in the Pacific Ocean and the Bahamas in the Caribbean.

Arctic Circle The line of latitude at 66° 30'N. All points along this latitude have a 24-hour period of continuous daylight in June and a 24-hour period of continuous darkness in December. Moving in a northerly direction from the Arctic Circle, the periods of continuous daylight or darkness become progressively longer.

arcuate delta SEE **delta**

arenaceous SEE **rudaceous rock**

arête A narrow knife-edged ridge which separates two corries, or a **corrie** and a **U-shaped valley**, formed by glacial erosion and **freeze-thaw** activity. In the Alps and Himalayas, some arêtes are only a few inches wide and may have steep slopes which drop more than 1,000 ft [300 m].

argillaceous SEE **rudaceous rock**

aridity Dryness because of lack of rainfall. An arid area receives less than 10 in [250 mm] of rainfall per annum, and only specialized drought-resistant plants are able to survive. There is much bare rock, with cactus, spiny bushes or tussocks of grass forming a sparse cover of vegetation.

arithmetic growth A growth rate with a regular progression; for example, 2, 4, 6, 8, 10. **Thomas Malthus** thought that food production would increase at an arithmetic rate, while population would grow at a **geometric rate**.

arithmetic mean An average figure for a set of data, found by adding up all the values of the set and dividing the total by the number of values. The figure may be distorted from a true average if the data set has a high proportion of very high or very low values. SEE ALSO **average**, **median**

Armorican SEE **Hercynian**

artesian A term denoting water which has moved from its place of origin by traveling underground, usually by means of percolating along an **aquifer**.

Structure of the Atmosphere

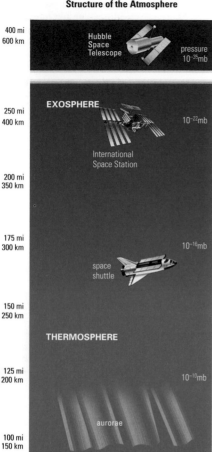

400 mi
600 km

Hubble
Space
Telescope

pressure
10^{-35}mb

EXOSPHERE

250 mi
400 km

10^{-22}mb

International
Space Station

200 mi
350 km

175 mi
300 km

10^{-16}mb

space
shuttle

150 mi
250 km

THERMOSPHERE

125 mi
200 km

10^{-10}mb

aurorae

100 mi
150 km

MESOSPHERE

60 km
100 km

10^{-3}mb

meteor trails

30 mi
50 km

ozone layer

STRATOSPHERE

6 mi
10 km

Concorde

10^{3}mb

TROPOSPHERE

artesian basin A synclinal structure with a **permeable** stratum between two **impermeable** strata. The largest artesian basin in the world is in Australia, and many farmers in southwest Queensland depend on artesian water for their cattle and their sheep. The water which comes up their boreholes probably fell as rain on the Great Dividing Range, several hundred miles to the east and northeast. In some cases, where the **water table** of the nearby hills is higher than the top of the well which is tapping the artesian water, the water may squirt out under its own pressure, possibly as a fountain.

arroyo SEE **wadi**

artificial fiber SEE **synthetic fiber**

ASEAN SEE **Association of Southeast Asian Nations**

ash The fine particles of rock thrown out by a volcanic eruption. If the ash is not blown very high, it will fall quickly and form a depositional area downwind of the **volcano**. If it is blown thousands of feet into the atmosphere, as at Krakatoa (Indonesia) in 1883, some of the dust may circle the Earth for several months. After the eruption of Mt. St. Helens, Washington, in 1980, a cloud of ash drifted eastward across the United States and the Atlantic Ocean. More recently, Mt. Pinatubo, in the Philippines, erupted in June 1991, after lying dormant for more than 600 years, thereby sending a huge amount of ash into the air.

aspect The direction a location is facing. The aspect is sometimes important for crop growing, as many plants need to face south in the northern hemisphere in order to get enough sunshine to ripen; for exam-

◄ **atmosphere** The main layers of the atmosphere, with approximate heights in miles (mi) and kilometers (km). The approximate heights reached by some vehicles are shown. The atmospheric pressure is given in millibars (mb). The pressure in the mesosphere is one millionth or less of that in the troposphere.

ple, grapes. It may also be important for people, for example, in order to avoid cold northerly winds.

assembly industry An industry engaged in gathering components and putting them together to make a finished product. The components may be made by several different firms. The automobile industry assembles the body, seats, windshield, engine, etc. as the car slowly moves along a conveyor belt.

Association of Southeast Asian Nations (ASEAN) An organization that was founded in Thailand in 1967 by Brunei, Indonesia, Malaysia, Philippines, Singapore and Thailand. It aims to help the development of its member states, especially in economic, social and cultural fields, and to promote stability and cooperation in the region. Burma and Laos joined in 1997, and Cambodia joined in 1999.

asymmetric fold A lop-sided fold that is caused by greater pressure from one side than the other. Most folds are asymmetric.

Atlantic coast SEE **discordant coastline**

atmosphere The layer of air which surrounds the Earth, acting as a shield against incoming objects, a radiation deflector, a thermal blanket, and a source of chemical energy for the Earth's inhabitants. It consists of the permanent gases, including oxygen, nitrogen, argon and helium, and the variable gases, such as **ozone** (which occurs mostly over 20 miles to 40 miles [30 km to 60 km] above the Earth's surface), water vapor (which is mostly less than 10 miles [15 km] above the Earth's surface), sulfur dioxide and carbon dioxide. About five-sixths of its mass is found in the first 10 miles [15 km], the troposphere. As the height above the Earth's surface increases, the density of the atmosphere becomes thinner, or more rare. However, the relative proportions of the gases remain fairly constant, although the amount of water vapor does decrease with height. Most of the weather phenomena occur in the lowest parts of the atmosphere. SEE ALSO **stratosphere, tropopause, troposphere, ionosphere, exosphere**

atmospheric pressure The pressure on the Earth's surface exerted by the weight of the **atmosphere**. This decreases with increasing altitude and is measured in millibars using a **barometer**. SEE ALSO **anticyclone, depression**

atoll A circular or horseshoe-shaped **coral reef** enclosing a central **lagoon**. It

▼ **atoll** An atoll begins as coral forming around a new volcano. When the volcano ceases to be active, it erodes away, leaving only the coral reef, which may by then be populated with plants, animals, and people.

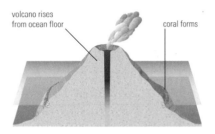

volcano rises from ocean floor

coral forms

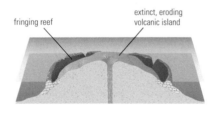

fringing reef

extinct, eroding volcanic island

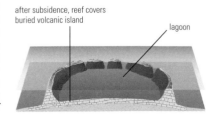

after subsidence, reef covers buried volcanic island

lagoon

begins as a bare volcanic peak, thrusting above the ocean surface. A colony of **coral** begins to form in the shallow water around the peak. When its sea-floor eruption is over, the **volcano** slowly sinks and is eroded, leaving the coral forming a ring (or partial ring) around its remnant. In time, all traces of the volcano may disappear and the coral reef will be left surrounding a lagoon. Atols are most common in the tropical waters of the Pacific Ocean and the Caribbean Sea.

attrition A process of **erosion** in which the particles or tools of erosion erode one another. For example, the sand particles in a river's load may gradually erode one another; small particles carried by the wind will also suffer from attrition, as will the pebbles on a beach. Attrition gradually reduces the size of the particles, making them smooth and rounded.

automation The use of machines to replace manpower. Automation is found in many manufacturing industries, such as automobiles or textiles, but usually only in the wealthier countries, since it is extremely **capital intensive**.

autumnal equinox SEE **equinox**

avalanche A large amount of snow and ice or rock falling suddenly down a mountainside. Avalanches occur when the temperature rises quickly, causing the snow to melt; they can also be started by a sudden loud noise.

average A measure of central tendency, i.e. the value that occurs most frequently, and is therefore most representative, of a set of values. ALSO CALLED **arithmetic mean**. SEE ALSO **median**

axis of the Earth A line joining the North and South Poles, around which the Earth rotates once every 24 hours.

azonal soil A soil associated with a particular environment, such as a scree slope, sand dune, or glacial moraine, on which little real soil development has taken place and there are no true soil horizons.

B

backing A meteorological term denoting the counterclockwise change of direction of a wind, such as southwesterly to southerly to southeasterly. SEE ALSO **veering**

backwash The downshore movement of water on a beach. When a wave breaks, the swash takes the water up the beach; the backwash brings it down again. The backwash may be on the surface or in the sand or shingle on the beach. It normally runs down the steepest slope, which is usually at right angles to the sea.

bacteria Part of a large group of microscopic unicellular or multicellular organisms. Bacterium live in dense numbers in favorable habitats and affect human beings both positively (such as in the breakdown of organic matter in soil) and negatively (for example, they may be the cause of serious diseases, such as typhoid or pneumonia).

badlands An eroded landscape in an arid or semiarid area. Because of the lack of veg-etation, the rainwater runs off very quickly and erodes any soft or exposed rocks, forming a steep and furrowed landscape which is impressive but quite useless for agriculture. The most noted badlands are found in South Dakota and Nebraska in the USA.

balance of payments A record, usually in yearly or monthly terms, of the difference between the income from **exports** and the cost of **imports** in a country's economy. Ideally, the total expended on imports will balance with the total received from exports, but this happens very rarely.

barchan I A crescent-shaped sand dune with the horns pointing downwind, found in parts of most of the large sand deserts of the world, such as southern California in the USA, the Sahara in Africa, and Turkestan in Central Asia. Barchans form in areas where the wind direction is fairly constant. Wind blows over the crest and around the edges of the horns, moving

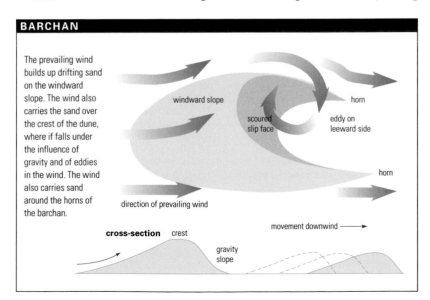

BARCHAN

The prevailing wind builds up drifting sand on the windward slope. The wind also carries the sand over the crest of the dune, where if falls under the influence of gravity and of eddies in the wind. The wind also carries sand around the horns of the barchan.

windward slope

scoured slip face

horn

eddy on leeward side

horn

direction of prevailing wind

movement downwind ⟶

cross-section crest

gravity slope

sand, so that the entire dune gradually moves downwind. **2** A similar feature of dry powdery snow. ALSO CALLED **barkhan**

bar graph A graph in which columns are used to represent the values of the information recorded. Several bars can be drawn close together to enable comparisons to be made. A simple bar shows a total value, while a compound bar can be subdivided to show constituents as well as total values. The divided bar graph can be used to show the kind of information which is often illustrated by a **pie chart**, and the divided bars are easier to draw than circles. The bars may be drawn horizontally or vertically. Horizontal bars can be placed one on top of the other to form a pyramid, as in a **population pyramid**; alternatively, they can be drawn side by side in a vertical arrangement, such as is often used to show monthly rainfall totals. ALSO CALLED **battleship diagram**

barkhan SEE **barchan**

barograph An instrument which writes a continuous record of **atmospheric pressure**. It consists of a small **aneroid barometer** with an arm, which moves up and down with the changes in pressure. At the end of the arm is a nib, which rests on a clockwork drum bearing a piece of special recording paper; as the drum revolves, a thin line is drawn on the paper. At the end of each week, the paper can be removed and a full record of the pressure changes can be seen. The barograph is more useful for recording the changes in pressure than for giving an accurate reading of atmospheric pressure at any particular time.

barometer An instrument for measuring **atmospheric pressure**. There are two main types: the mercury barometer and the **aneroid barometer**, which is the less accurate of the two but is the type normally used in **barographs** and found in the home. The mercury barometer consists of a glass tube containing mercury, which is inverted over a small reservoir of mercury; as the pressure rises, the weight of the atmosphere forces mercury out of the reservoir and up

the tube. The tube has a scale on it, so that the pressure can be read off at the top of the column of mercury. As the pressure falls, the column of mercury gradually moves down. A column of mercury 30 in [760 mm] in length represents the average pressure at sea level, and so the mercury barometer has to be quite a long instrument in order to contain such a long tube.

barrage A large structure of concrete, or earth and rocks, designed to block the flow of a river in order to form a lake. The lake may be used as a source of water for **irrigation**, or may be designed to reduce flood danger. There is a barrage across the Rance estuary near St. Malo in France, where a tidal power station has been erected. SEE ALSO **dam**

barrier reef A **coral reef** set away from the coastline but parallel to it, with a deep **lagoon** separating the two. The best example is the Great Barrier Reef, off the coast of Queensland, Australia, which stretches for over 1,000 miles [1,600 km].

barrow SEE **tumulus**

basalt A black or dark colored **igneous** rock which consists of fine grained minerals because it was formed near the Earth's surface from lava that cooled fairly quickly. The dark color of basalt is caused by the presence of minerals which are basic rather than acidic in character, such as augite, hornblende and **mica**. Basalt sometimes forms hexagonal columns when it cools from the liquid lava, as in Giant's Causeway in Northern Ireland, Devil's Postpile National Monument in California, USA, and at Fingal's Cave on the island of Staffa, off the coast of Scotland. Basalt is formed by extrusive volcanic activity from lava that is fairly free-flowing and liquid. Flows of basalt sometimes spread over large areas of countryside; for example, in Iceland, the Snake/Columbia plateau in the northwest USA the Deccan plateau in India, and.

basic lava Free-flowing lava, spreading quickly across the countryside after an

BAR GRAPH

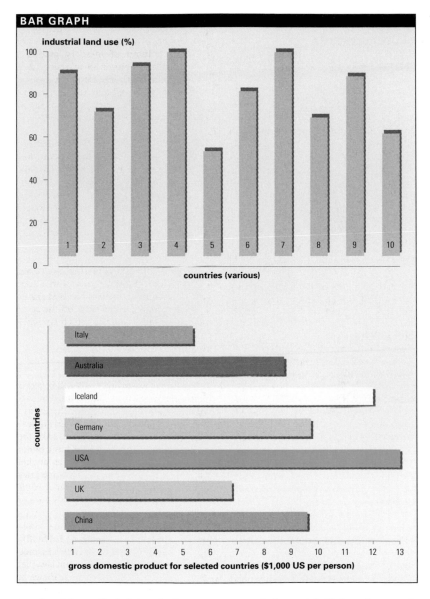

industrial land use (%)

countries (various)

gross domestic product for selected countries ($1,000 US per person)

eruption. It is usually dark in color because it contains basic ferromagnesian minerals and is less than 50% silica. It may erupt through craters or from fissures, and forms **basalt** when it solidifies; fissure eruptions are likely to form plateaus. If a volcanic mountain forms, it is likely to have gently sloping sides which spread over a very large area; for example, Mauna Loa in Hawaii, USA. Basic lavas are common in Iceland on the Mid-Atlantic Ridge; old basic lavas are found in the Antrim Mountains of

BAROMETER

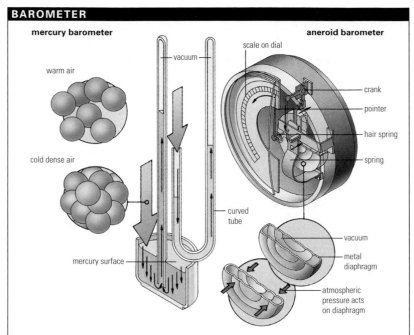

mercury barometer

aneroid barometer

warm air

vacuum

scale on dial

crank

pointer

hair spring

spring

cold dense air

curved tube

vacuum

metal diaphragm

atmospheric pressure acts on diaphragm

mercury surface

Barometers are instruments used to measure air pressure. A glass tube, open at one end, is filled with mercury (which is used because of its density) and inverted in a bowl of mercury. The column of mercury drops, creating a vacuum above it, until its weight is balanced by the weight of the air pressing on the mercury surface. The column of mercury is 760mm tall at sea level and 45° latitude. The bowl

of mercury can be replaced by using a curved tube. Warmer, less dense air will exert less pressure than cold dense air. The aneroid barometer dispenses with the mercury, changes in air pressure being registered by the movement of a flexible metal diaphragm (the "bellows") containing a vacuum , which is connected by a series of levers to a sprung indicator pointer and scale.

Northern Ireland and on the island of Mull in western Scotland. **Weathering** gradually breaks up the lava to form soil which will be rich and productive, as in Java, Indonesia, or on the slopes of Mt. Etna in Sicily. SEE ALSO **acid lava**, **viscous lava**

basin I An area in which the strata dip from the outside edges toward a central point, such as the Donets and Kuznetsk basins in Russia. **2a** The catchment area of a river. **2b** An area of inland drainage in desert regions, where rivers which are unable to reach the sea either flow into lakes or evaporate in salt flats, such as the Great Salt Lake in Utah, USA, and Lake Eyre in Australia.

3 A large hollow or depression in the Earth's surface, partly or wholly surrounded by higher ground. **4** A type of irrigation, found in India, Egypt and elsewhere, in which embankments are constructed to prevent the escape of flood or irrigation water from a field. The water gradually sinks into the soil, thus enabling crops to be grown.

batholith A large area of **igneous** rock which cooled and solidified below the surface. Because of this it contains large crystals and will probably be **granite**. Batholiths can be seen at the surface when the overlying rocks have been worn away by **erosion**. As they are quite hard rocks, they generally form high ground, and sometimes they can form large mountain ranges, such as the Sierra Nevada in California, USA.

battery farming SEE **factory farming**

battleship diagram SEE **bar graph**

bauxite The ore from which aluminum is obtained. Formed in tropical regions, there are rich deposits of bauxite in Guinea, Jamaica, the USA, Brazil and Surinam.

bay An indentation or inlet in the shore of a sea or lake. It is generally formed as a result of varying rates of **erosion**, the softer rocks having worn away faster than neighboring harder rocks. Bays may provide useful sheltered anchorage for boats and are often the site of settlements.

bay bar A ridge of sediment deposited across the mouth of a bay and attached to the land at both ends. It consists of material which was deposited by a river or car-ried along the coast by **longshore drift**. If it completely blocks the bay from the sea, deposits will pile up on the landward side of the bar. This will gradually form a **delta** of reclaimed land. SEE ALSO **spit**

beach The land on a shore between high-water mark and low-water mark. Some beaches consist of smoothed rocks which have been eroded by the sea; others may contain large and rugged rocks, also resulting from **erosion**. Many beaches contain material of deposition, such as shingle, sand or mud. The constituents of a beach will depend on the coastal rock types, the amount of erosion, neighboring rock types along the coast, and the amount of **longshore drift**; also, the presence of **groynes** may have a significant influence on the nature of a beach. Constructive action along a beach may be caused by the swash; destructive action is associated with the **backwash**.

bearing A **compass** point measured in degrees from 0 to 360. To work out a bear-

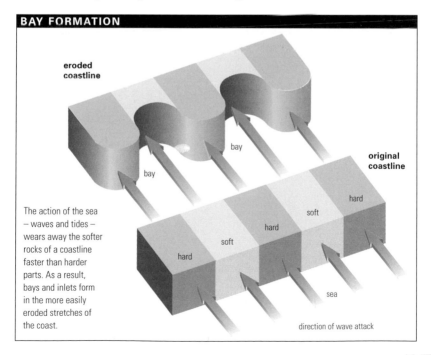

BAY FORMATION

eroded coastline

bay

bay

original coastline

hard

soft

hard

soft

hard

soft

hard

sea

The action of the sea – waves and tides – wears away the softer rocks of a coastline faster than harder parts. As a result, bays and inlets form in the more easily eroded stretches of the coast.

direction of wave attack

BEAUFORT SCALE			
Scale Number	Description	Wind Speed (km/h)	(mph)
0	Calm	0	0
1	Light air	1 – 5	1 – 3
2	Light breeze	6 – 11	4 – 7
3	Gentle breeze	12 – 19	8 – 12
4	Moderate breeze	20 – 28	13 – 18
5	Fresh breeze	29 – 38	19 – 24
6	Strong breeze	39 – 49	25 – 31
7	Near gale	50 – 61	32 – 38
8	Gale	62 – 74	39 – 46
9	Strong gale	75 – 88	47 – 54
10	Storm	89 – 102	55 – 63
11	Violent storm	103 – 117	64 – 72
12	Hurricane	118+	73+

ing start from north, which is 0° and 360°, and progress in a clockwise direction; east is 90°, south is 180°, and southwest is 225°.

Beaufort scale A scale of wind speed devised by Admiral Beaufort at the beginning of the 19th century. The Beaufort scale is used internationally. It ranges from 0, where wind is calm (smoke rises vertically), to 12, denoting a hurricane with wind speeds of more than 73 mph [118 km/h]. SEE TABLE

bedding plane The dividing line between two different strata of **sedimentary rock**. A bedding plane generally indicates a slight change of rock type; each stratum would have been deposited over a period of years, and then consolidated to form a layer of rock.

bedrock The solid rock which lies below the soil and the **regolith** layers. In soil profiles it will be the C or D horizon. In some places, such as the North American prairies or East Anglia in England, the bedrock is several feet below the surface, but in other places, especially on hillsides, it may be within a few inches of the surface.

beltway A road which goes around a city in order to divert traffic away from the inner

city and reduce congestion. Many cities have ring roads.

Benguela Current A cold **ocean current** that is found off the southwestern coast of Africa.

bergschrund A large **crevasse** which forms in the ice at the upper end of a **corrie**, near the back wall. It is caused by the downhill movement of the ice, which gradually pulls away from the back wall, opening up a crack or crevasse.

berm A narrow horizontal ridge or shelf composed of beach debris from storm waves, that runs parallel to the shore, above the foreshore.

beta index A quantitative method of analyzing a network of communications. It involves counting up the number of edges (or arcs) and nodes (or vertices) in a network of roads or railroads. Dividing the number of nodes into the number of edges will give the beta index. The figure for one network can then be compared with the figure for another network in a precise and quantitative way. If there is one circuit in the network, the index will be one; two circuits will give an index of two, and so on. The greater the index, the better the connectivity of the network.

B horizon The layer beneath the A horizon in a fully-formed **soil profile**. It has less weathered material and less **humus** than the A horizon, but often contains chemicals which have been washed down from above. SEE ALSO **A horizon**, **C horizon**

bid rent theory The theory that rent or land values decrease with increasing distance from a central location. The center may be a small town or the center of a large city. Store and office owners are prepared to pay higher rents for central and accessible locations, and they therefore tend to occupy the most expensive central areas. The ideas of bid rent occur in the **von Thunen** land-use model, and also in land-use models of urban areas.

It is a sound general idea, but there are often exceptions caused by relief features, communications, or the idiosyncrasies of human beings.

bilharziasis SEE **schistosomiasis**

biogas The gas which is produced by rotting plant or animal material, such as manure; methane. It is a renewable source of energy and has already been used on a small scale to produce electricity in the USA and as a substitute for firewood in parts of India.

biogeography The study of the geographical distribution of plants and animals. Biogeography includes the study of **ecology** and **ecosystems**, soil studies and many other related topics.

biological weathering SEE **weathering**

biomass The total organic matter of the plants and animals in a given area. It is usu-ally expressed as the dried weight per unit of area, such as per square foot or square meter. Figures range from 10 lb per sq ft [45 kg per sq m] in tropical rainforests, to only 0.005 lb [0.02 kg] in polar regions. The biomass of temperate deciduous forests is 6 lb per sq ft [30 kg per sq m]; savanna areas average 1.2 lb [6 kg]; deserts are 0.15 lb [0.7 kg], and cultivated land is about 0.2 lb per sq ft [1 kg per sq m]. The animal proportion of the biomass is very small compared with that of the plants.

biome A major type of **environment**, such as savanna, tropical rain forest or tundra, considered as a whole with its plant and animal life.

biosphere The area of the Earth's crust and atmosphere where life is found. This zone reaches some 10 ft [3 m] below ground and 100 ft [30 m] above it; it also includes water regions, stretching 640 ft [200 m] deep, where most marine and freshwater life is found.

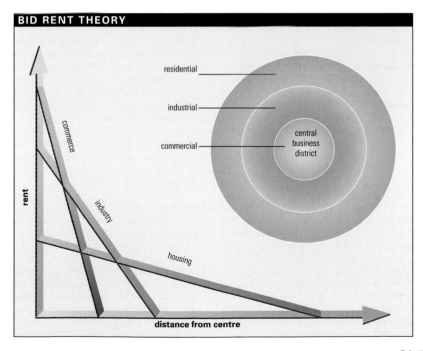

BID RENT THEORY

bird's-foot delta SEE **delta**

birth control The prevention of pregnancies by contraception, sterilization and abortion. Some countries, for example China, have strict birth-control policies in an attempt to limit family size and thereby slow down a fast-growing population.

birth rate The number of live births per year for every thousand of the population under consideration. A birth rate of 40 per thousand means that for every 1,000 people in the population there will be an average of 40 live births each year. Some **Third World** countries, such as Kenya, still have high birth rates, but in most Western countries the figure is now much lower, about 12 to 15. ALSO CALLED **crude birth rate**

bituminous coal A type of coal which contains about 80% carbon. It burns quite well, but gives off some smoke and leaves some ash. The major users are domestic households and thermal power stations; for example, bituminous coal from western Pennsylvania, USA, is used by electric utility companies. SEE ALSO **anthracite, brown coal**

blackband iron ore A thin seam of iron ore, dark in color, found in **coal measures**. The blackband deposits found in such areas as South Wales, central Scotland and the Ruhr valley helped the iron and steel industries to develop in those areas, although the deposits did not last very long and steel works were relocated nearer ports to use imported ore.

black earth SEE **chernozem**

black ice A layer of transparent ice that is often invisible, thereby making it very dangerous when it forms on roads.

blanket bog A **bog** that completely covers an area of usually horizontal land. Its formation is dependent on high humidity and rainfall, and only steep slopes and rock outcrops will remain dry. Commonly found in Newfoundland and Ireland.

blizzard A strong and very cold wind containing particles of ice and snow that have been whipped up from the ground.

block mountain A mountain or plateau uplifted between two or more parallel faults. Good examples can be found in the Sierra Nevada in the USA, Meseta in Spain, the Black Forest in Germany, the Vosges in France and the Grampians in Scotland. They are all steep sided because of the effects of the faulting. ALSO CALLED **horst**

blowhole A hole in a cave or cliff through which water is forced, especially at high tide. Blowholes are formed by **erosion**, probably along a **joint** or line of weakness in the rocks, and commonly occur in cave roofs. Examples can be seen on many coastlines, such as in the Caymen Islands or Caithness in northern Scotland.

blue-collar worker A manual worker; an employee who works in a potentially dirty environment and would therefore need to wear overalls or other protective clothing. SEE ALSO **white-collar worker**

bluff A steep slope at the side of a river formed by erosion on the outside of a **meander**. Bluffs may be quite small, but some are 300 ft [100 m] or more in height. Since they are steep, they are often covered with trees and are not used for farming. ALSO CALLED **river cliff**

bocage A type of landscape found in the Brittany and Normandy regions of western France. It consists of small farms in fields surrounded by wooded hedges.

bog An area of poor drainage that causes the partially decomposed vegetation to become wet and spongy, thereby forming an acid peat. SEE ALSO **blanket bog**

bora A cold wind which blows from the north or northeast into the northern part of the Adriatic. It is associated with the movement of a low-pressure system along the Mediterranean, into which the air is drawn. The bora consists of cold air blowing out from an **anticyclone** over Europe, and is similar to the **mistral** of the Rhône valley.

boulder clay SEE **till**

bourne A stream, especially on chalklands in southern England. Many bournes dry up for part of the year, normally in summer, when the **water table** falls.

braided stream A stream or river which splits into two or more smaller channels that rejoin further downstream. The split or braiding is generally caused by deposits of sand or mud which have been dumped by the river. At times of flood the deposits may build up above the normal river level, and then vegetation may grow on them. When this happens, the small muddy islands become permanent. Braiding also occurs where streams are flowing out from melting ice and there are vast deposits

of sand and gravel, through which the river finds several small channels.

Brandt Commission (1977) An international commission chaired by Willy Brandt, the former chancellor of West Germany, which studied the problems facing the **Third World**. The commission introduced the terms "**the North**" and "**the South**" to refer collectively to the more advanced countries and the less developed countries, and suggested ways in which improvements could be made.

break of bulk The unloading and breaking down of a ship's bulk cargo for transportation by rail or road. Many ports have become break-of-bulk locations, which have also attracted industrial development because of the availability of **raw materials**. The term "break of bulk" is now applied to any change from one form of transportation to another.

breccia A type of rock consisting of angular fragments of other rocks which have been joined together by a fine-grained

▼ **blowhole** A blowhole forms as the sea erodes a cliff. First a cave appears. Then, though a combination of erosion and roof collapse in the cave, a blowhole develops, connecting the sea and the land's surface.

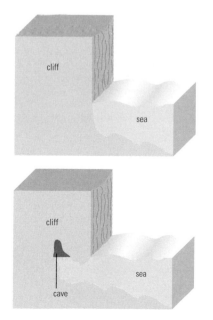

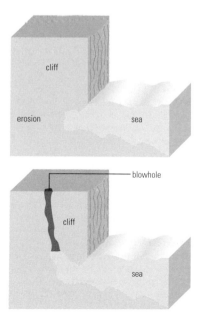

cementing material such as clay. The rock fragments are often of desert origin, but some come from volcanic eruptions.

Breckland An area of sandy heathland on the Norfolk–Suffolk border in East Anglia, England. Laid down by an **ice sheet**, it is an expanse of sandy, infertile **till**, which is quite different from most of East Anglia, where the glacial deposits have formed many areas of rich soil. Much of Breckland has now been planted by the Forestry Commission, but some has been utilized as farmland, thanks to the addition of fertilizers.

brickearth SEE **loess**

bridging point A point where a bridge has been built across a river. It may be where the river narrows, where there is firm ground alongside the river for good foundations, or where there is an island in the middle of the river; the presence of an island always makes bridge building a little easier, as, for example, in Paris, France, and Montreal, Canada. The lowest bridging point is the last bridge downstream. As this is also the first bridge upstream, it will possibly be the highest point to which boats can travel, unless the bridge can be opened to allow boats to pass through (for example, Tower Bridge on the River Thames in London, England).

broad A small shallow lake in East Anglia, England, formed as a result of **peat** digging in the Middle Ages. In 1989 the Norfolk Broads were designated as a National Park.

broadleaf evergreen tree Any broadleaf tree, as opposed to a conifer, which retains green leaves all the year around. Most broadleaf trees, for example, lime and ash, are deciduous and shed their leaves for the winter. However, there are some broadleaf trees which remain evergreen because of heat and moisture all year around, as in the Amazon basin in Brazil. A more temperate example of this can be seen along the southern coastal regions of the Atlantic states of the USA.

Bronze Age A period in human development when bronze (an alloy of copper and tin) was used to make weapons and tools. This occurred around 1500 BC and also saw the development of the wheel, the domestication of animals, and the use of writing and arithmetic.

brown coal A type of coal, only 50% of which is carbon. Of poor quality, it burns with much smoke and leaves large amounts of ash, and so is often converted into electricity at the mining location to save transportation costs. Brown coal is an important source of energy in parts of Russia, Poland and Germany. There is also a large mining area in southeast Australia. ALSO CALLED **lignite**. SEE ALSO **anthracite, bituminous coal**

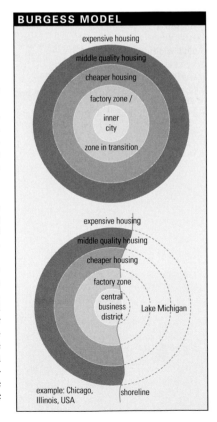

BURGESS MODEL

expensive housing
middle quality housing
cheaper housing
factory zone /
inner city
zone in transition

expensive housing
middle quality housing
cheaper housing
factory zone
central business district
Lake Michigan

example: Chicago, Illinois, USA
shoreline

Burgess model A concept of urban layout, as defined by the American, E.W. Burgess. In the 1920s, he studied Chicago in the USA and realized that cities often grow in a series of concentric rings. His five major zones were: **1** the **central business district (CBD)**; **2** the transition zone, where factories were located; **3** a zone of cheaper housing where the factory workers would live; **4** residential homes, which were more expensive; **5** the commuter zone, where the most expensive houses could be found, and the inhabitants could afford to travel into the city each day. His rather simple model has been criticized by many authorities, with justification, but it is surprising how many cities have some concentricity in their land-use zones. It is also known as the **concentric ring theory**.

bush In Australia, New Zealand, the USA and parts of southern Africa, an area of scrub where few people live and the landscape is still wild and isolated.

bush fallow SEE **shifting cultivation**

bustee An area of shanty development in Calcutta, India. Thousands of people sleep on the sidewalks, but hundreds of thousands live in simple, primitive accommodation. Huts made of cardboard, leaves and branches or even large drainage pipes are used.

butte A small, isolated flat-topped hill which is surrounded by deeply cut river valleys. A stratum of hard rock forms the top of the butte, and this protects the underlying layers from **erosion**. Formerly part of a larger plateau, it is a feature which has been cut off and isolated by river erosion. The best examples can be seen in the southwestern USA. SEE ALSO **mesa**

Buys Ballot's Law The principle, observed in 1857 by Danish meteorologist C.H.D. Buys Ballot, that if an observer in the northern hemisphere stands with his or her back to the wind, the **atmospheric pressure** will be lower to the left than to the right, the reverse being true in the southern hemisphere. This is caused by the fact that winds blow from high pressure to low pressure, and are deflected to the right in the northern hemisphere but to the left in the southern hemisphere.

C

caatinga The thorny scrubland found in the interior of northeast Brazil. The vegetation, which includes cactuses and acacias, is drought resistant (**xerophytic**) because it has to be able to withstand prolonged dry spells, especially in the winter months.

cacao A small evergreen tree, *Theobroma cacao*, grown for its seeds, which are roasted and ground to produce cocoa. Found in equatorial areas, it needs a constant temperature of about 77°F [25°C] and rainfall throughout the year; strong winds can be harmful. Brown seed pods, which contain the cocoa beans, grow out of the trunk and main branches. Among the world's principal cocoa producers are Ivory Coast, Brazil, Ghana, Malaysia, Indonesia and Ecuador.

Cainozoic SEE Cenozoic

calcareous The term used to denote a soil or rock containing calcium carbonate ($CaCO_3$). Chalk and limestone are calcareous rocks, though limestone also contains many other minerals. Soils which develop on such rocks are calcareous. SEE ALSO **rendzina**

calcite The mineral which is the main component of **calcareous** rocks; calcium carbonate.

caldera A large crater formed by the collapse of the cone of a **volcano**. Subsidiary cones may form within a caldera as a result of subsequent eruptions. Crater Lake in Oregon, USA, is a large lake which has formed in a caldera.

Caledonian folding A major period of **earth movements** to form fold mountains, which occurred near the end of the Silurian period 400 million years ago. During this episode, the Scottish Highlands and the mountains of Norway and Sweden were formed.

calorie The unit of heat energy, defined as equal to 4.18 joules, required to raise the temperature of 1 gram of water from 15°C to 16°C [59°F to 61°F]. The "kilocalorie" is 1,000 calories and is used to describe the calorific value of food (though the prefix "kilo" is usually omitted). This energy can be generated by food and can be related to the amount of food required to keep a healthy human being fit and active. Starch, fat and sugar generally give more calories than lean meat, fruit and vegetables. It is thought that about 2,500 calories would be adequate for each day, and 2,000 would be the absolute minimum. An energetic person or a manual worker might require 3,000 calories or more. Probably 60% to 70% of the world's inhabitants obtain 2,000 calories or less, on an average day, and are suffering from hunger or malnutrition. SEE ALSO **joule**

Cambrian A geological period that began about 570 million years ago and ended about 500 million years ago. It is the first division of the **Paleozoic** era, and many rocks in North Wales are from this period. Similar rocks are found in the northeast USA, which at that time was part of the same land mass as Wales. SEE ALSO **geological column, geological time scale**

canal A man-made waterway used for inland transportation and irrigation. Canals were especially popular in the late 18th century due to the poor quality of the roads and the cheapness of water transportation.

canyon A deep river valley with steep sides. Downward **erosion** by the river has not been accompanied by rainwash erosion of the valley sides, since the harder layers of horizontally bedded strata protect the softer layers. Large canyons are found in the southwest USA, notably the Grand Canyon of the Colorado River, which is nearly 1.25 miles [2 km] deep in places.

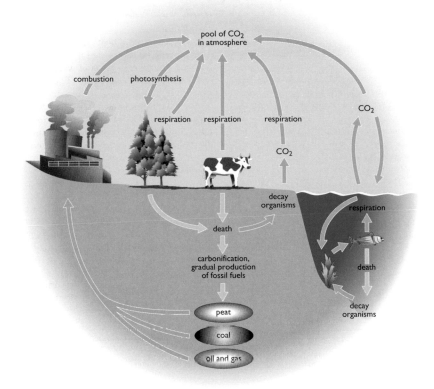

▲ **carbon cycle** Elemental carbon is in constant flux. Gaseous carbon dioxide (CO_2) is first incorporated into simple sugars by photosynthesis in green plants. These sugars may be broken down (respired) to provide energy, a process that releases CO_2 back into the atmosphere. Alternatively, animals that eat the plants also metabolize the sugars and release CO_2 in the process. Geological processes also affect the Earth's carbon balance, with carbon being removed from the cycle when it is accumulated within fossil fuels such as coal, oil and gas. Conversely, large amounts of carbon dioxide are released into the atmosphere when such fuels are burned.

CAP SEE **Common Agricultural Policy**

cap rock A layer of rock on the top of a plateau which is hard and resistant to **erosion**. The term may also apply to a hard stratum at the top of a waterfall as, for example, at the Niagara Falls in Canada/USA, or on the top of hills as, for example, on the North Downs in southeast England where the clay is topped with flints.

capillary action In soil, the tiny spaces ("capillaries") between soil particles attract moisture and the surface tension is strong enough to cause water to move up through the soil, above the **water table**.

capital 1 The main town in a "province" or country, that is often the seat of government. **2** An economic term used to define assets with a monetary value. Capital is one of the factors of production and includes plant, machinery and buildings as "fixed capital," and raw materials and fuel as "floating capital." "Finance capital" is the actual money needed for the production of goods.

capital intensive The term used to denote an industry or business activity in which a relatively large proportion of money is invested in equipment or machinery, as opposed to labor. For example, highly automated car factories or cereal farming using combine harvesters are very capital intensive. SEE ALSO **labor intensive**

capitalism An economic system characterized by the private ownership of, and private investment in, the production of goods. It involves the idea of a **free market** – where people are rewarded in proportion to their economic contributions. SEE ALSO **communism**

carbonation The process by which rainwater containing carbon dioxide from the **atmosphere** causes chemical weathering of **calcareous** rocks by dissolving the calcium carbonate. The dissolved materials are then carried away in the water, but

may be deposited elsewhere to form **stalagmites** or **stalactites**.

Carboniferous A geological period, part of the **Paleozoic** era, lasting from 345 to 280 million years ago. It is often called the "Age of Coal" because of its extensive swampy forests of conifers and tree ferns that turned into most of today's **coal** deposits. Amphibians flourished, marine life abounded in warm, inland seas, and the first reptiles appeared. SEE ALSO **geological column, geological time scale**

cartography The study and production of maps and charts.

cash crop A crop which is grown for sale to earn money, rather than for consumption by the farmer and his or her family. In the more advanced countries it is likely that the farmers will sell everything they produce, as, for example, the cereal growers

▶ **cave** Caves form in various ways, but the most common way, which produces the largest caves, is when water dissolves soluble rock such as limestone. The process of dissolving is known as chemical weathering. In the illustration, surface waters percolate through tiny fissures (1) in the limestone, widening these channels before they reach the water table – the upper surface of the water-saturated part of the ground (2). The waters then flow horizontally towards a natural outlet – in this case a river (3) – dissolving away limestone in their path. The underground waters eventually carve out a

main horizontal channel (4) at the depth of the water table. As it widens, this channel draws an increasing volume of water, thus accelerating its growth. Some of the vertical shafts also begin to attract more than their fair share of surface drainage, and may develop large funnel-shaped hollows, or sink holes, around their mouths (5). On the

surface, the river cuts through the limestone, and the water table drops (6). The water in the main underground channel drains out, seeking new paths to the water table, and the process of channel-carving begins again at the lower level. Wherever other rocks cap the limestone, such as

sandstone (7) and impermeable clays (8), water may be trapped on the surface in hollows and depressions (9), or may give rise to rivers that run along the surface, often high above the river that defines the water table (10).`

on the prairies of Canada and the USA. For a comparison, see **subsistence crop**.

cataract SEE **rapids**

catch crop A crop which is grown just after a main crop has been harvested. It is likely to be a crop which can be grown and harvested in a relatively short time. The cultivation of catch crops is a way of making full use of the land; it is a source of extra income for the farmer and also ensures that the soil is not left exposed.

catchment area 1 The area which is drained by a river and its tributaries; the area from which a river catches or collects its water. A catchment area can be very small – just a few square feet – but it may cover an enormous area, as, for example, the Amazon basin in South America. Apart from some loss by evaporation, all the water that falls as rain within a catchment area

will either run off on the surface or sink into the ground, eventually reaching the river which drains the basin. The speed at which the water reaches the river or its tributaries will depend on the vegetation cover and the rock type. If there are settlements in the area, the effects of man-made drains will also be relevant. ALSO CALLED **drainage basin**, **river basin** SEE ALSO **watershed 2** The area from which people may come to a central institution, such as a school or hospital. Large cities will have large catchment areas, whereas small towns are likely to have only a small catchment area.

cave A hollow space resulting from the **erosion** of rock. Caves are common in **limestone** areas where water dissolves the rock. Enlarged joints and bedding planes provide lines of weakness where erosion can set in. In many limestone areas there are underground streams, which contribute to the process of erosion and the formation of

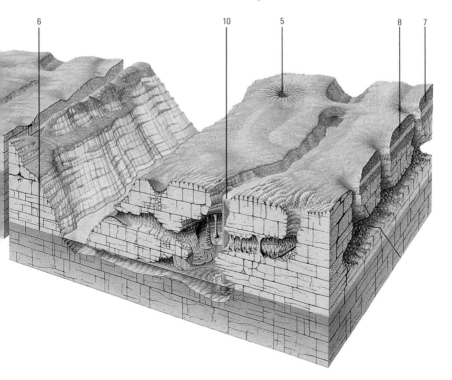

long networks of caves, caverns and tunnels. Caves also form on coastlines, where the exposed rock of cliffs is attacked and eroded by the sea. **Corrasion** is caused by the presence of sand and pebbles, and, as well as corrasion of the rocks, there is likely to be **hydraulic action**. Air often becomes trapped in fissures in the rock; whenever a wave breaks, the air is compressed and this creates a hammering effect which helps to weaken the rock.

CBD SEE **central business district**

Celsius A widely used temperature scale devised by the Swedish astronomer Anders Celsius (1701–44). It is based on a 0 to 100 scale; water freezes at 0° and boils at 100°. ALSO CALLED **centigrade** SEE ALSO **Fahrenheit**

Cenozoic The most recent of the geological eras; the name means "the recent period of life." It dates from about 65 million years ago and is the time when mammals became an important group. The era contains the geological epochs of the Paleocene (oldest), Eocene, Oligocene, Miocene and Pliocene (youngest), which make up the **Tertiary** period; and the **Quaternary**, which is the **Ice Age** together with the 11,000 years since the last ice advance. During the Cenozoic there was considerable tectonic activity, which formed most of the present high mountain ranges, including the Alps, Himalayas, Andes, Rockies, Atlas and Caucasus. ALSO CALLED **Caenozoic, Cainozoic, Kainozoic**. SEE ALSO **geological column, geological time scale**

census A survey of a given population, mainly to find its size at a given time, but also to discover its nature; for example, age, occupation, marital status, etc.

centigrade SEE **Celsius**

central business district (CBD) The central part of a town or city, which is likely to contain many offices and business headquarters, as well as stores, banks and,

possibly, council offices. Land values are usually very high, and only the larger and wealthier organizations are able to pay the high rents or rates. The main advantage of the central business district is its central location, which makes it the most accessible part of a town or city. The retail and business functions which take place there are all helped by this **accessibility**. The demand for land (therefore making it more expensive) explains why high-rise buildings and skyscrapers are commonly found in CBDs. There are sometimes underground car parks, and buildings may extend below the surface, as well as above, in order to make the fullest possible use of the available space.

central place 1 A settlement which is central to its surrounding area, and which provides goods or services to that area. **2** A store or group of stores, as, for example, in a suburban area, which represents a center to local inhabitants, though not necessarily to the town as a whole.

central place theory The theory that in an **isotropic** area settlements grow up at very regular distances, in order to serve the people who live in the surrounding rural region. It is associated with Walter Christaller, who in 1933 wrote about the settlements in his home area of southern Germany. His studies showed that there were large numbers of very small settlements, or hamlets, which he called first order settlements. Every few kilometers slightly larger settlements – villages or towns – would grow, and at even greater distances cities would develop. Each settlement would be surrounded by an area which it served – its **catchment area**. Circular catchment areas give overlaps as well as gaps, and so hexagons have been used instead to represent catchment areas. The hamlets would offer only a small range of services, but the villages (second-order settlements) would provide more stores. The towns (third-order settlements) would have many more services and facilities, as they would be serving a far larger catchment area. A map can be drawn to show the hexagonal catchment area and the settlement hierarchy.

centrifugal Movement away from a center.

centripetal Movement toward a center.

cereal Any of the cultivated grasses grown for their edible grain, such as barley, rye, wheat, rice and oats. Different cereals are grown in different environments according to the varying climatic conditions required for their growth; rice needs high temperatures (77°F [25°C]) and wet conditions during the early months of growth; barley and rye can both survive in areas with very cold winters, providing summer temperatures reach 59°F [15°C]. Barley grows well with an annual rainfall of only 22 in [550 mm], but oats can grow in far wetter areas, with as much as 60 in [1,500 mm] annually.

chalk A soft sedimentary rock, which is very light in color because it consists of the skeletal remains of dead sea animals. It is almost pure calcium carbonate in places and is a very **permeable** rock, which means that water can pass through quite easily. Although chalk is soft, because of its permeability it does not always wear away very quickly, and sometimes stands up in steep slopes and cliffs. There are many chalk cliffs on the south coast of England, for example, the "white cliffs of Dover." There is little surface water on chalklands, although there are many **dry valleys**. These are the relics of river systems which used to flow on the surface, either because the **water table** was then much higher, or because of **permafrost** conditions during a glacial phase of the **Ice Age**, when the frozen **subsoil** prevented the water from sinking beneath the surface.

channel Any waterway where water is free-flowing, either artificial (for example, a **canal**) or natural.

chaparral A low dense scrub consisting of small evergreen bushes that is found in the USA, especially California. It is similar to the **maquis** of the Mediterranean lands.

chemical weathering SEE **weathering**

chernozem A dark brown or blackish deep rich and fertile soil which occurs in areas of **temperate** grassland. The top layers of the soil are rich in humus, lime and most plant nutrients, and excellent crops of wheat have been grown for years on chernozems. Because the climatic conditions are quite dry, there is an upward movement of minerals through the soil. There are large areas of chernozem in the Ukraine and neighbouring parts of Romania and Hungary, and in the prairies of the USA and Canada. ALSO CALLED **black earth**

chert A rock made up of silica, probably derived from fragments of dead sea creatures. Chert occurs in chalky areas, but while chalk consists of the remains of sea creatures whose shells and skeletons were of calcium, chert is made up of those animals whose structural remains were of silica, such as sponges.

chestnut soil A zonal soil that is usually dark brown over a lighter colored soil. It is found in drier grasslands than those associated with **chernozems** and thereby contains less humus. For example, chestnut soils are found in parts of the Pampas in Argentina.

china clay A white clay formed from the disintegration of **granite**. Obtained from **feldspar**, which is one of the minerals that makes up granite, it is used in the manufacture of high-quality porcelain. It is also used medicinally and in the manufacture of certain types of paper. ALSO CALLED **kaolin**

chinook A warm wind which blows down the eastern side of the Rocky Mountains in Canada and the USA. It contains air which has come from the western side of the Rockies. As the air rises over the mountains, it is cooled quite slowly at the wet (or **saturated**) **adiabatic rate** – and rain or snow falls on the mountains, so that the air becomes progressively drier. When it descends the eastern side of the Rockies, it gains heat at the **dry adiabatic rate**, becoming a warm wind by the time it crosses the prairies. It is commonest in winter and spring, when depressions come inland from the Pacific Ocean. The warmth of the chinook is often sufficient to melt snow and help to thaw out the soil, enabling cereals to be planted in the spring. It has been known for the chinook to change the temperature by as much as 36°F [20°C] in about 15 minutes.

chitamene SEE **shifting cultivation**

C horizon The parent material from which a **soil** has been created; the **subsoil**. SEE ALSO **A horizon, B horizon**

choropleth map A map with shading to provide quantitative information about different areas or regions. Choropleth maps show the average information relating to a given area, usually an administrative district, because of the availability of information and data. It is normal to draw choropleths in black and white, with white indicating the lowest numbers and black the highest. Dots and cross-shading may also be used for different classes of information, and each area of shading will have a definite boundary line, not a transitional area. ALSO CALLED **shading map**

cirque An armchair-shaped hollow on a hillside, which has been cut by ice and **freeze-thaw** activity. If the hollow is deep enough, a lake will form in it when all the ice has melted. The sides and back wall of a cirque are steep and may be several hundred feet in height. Cirques occur in most high mountain regions, including the Alps, Andes and Rockies. ALSO CALLED **combe, corrie, cwm**

cirrocumulus A high cloud consisting of small patches of ice crystals often stretched out in a linear form. This type of cloud is called a "mackerel sky" or "buttermilk sky." SEE ALSO **cloud**

cirrostratus A thin sheet of cloud which consists of high-level accumulations of ice crystals. SEE ALSO **cloud**

CIRQUE

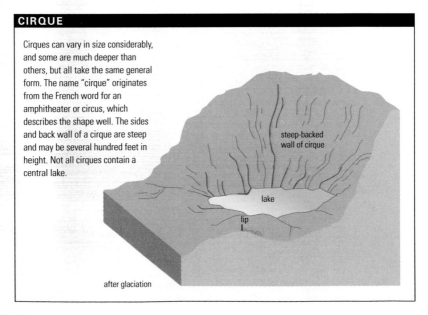

Cirques can vary in size considerably, and some are much deeper than others, but all take the same general form. The name "cirque" originates from the French word for an amphitheater or circus, which describes the shape well. The sides and back wall of a cirque are steep and may be several hundred feet in height. Not all cirques contain a central lake.

steep-backed wall of cirque

lake

lip

after glaciation

cirrus A type of cloud which forms at great altitude, is thin, wispy or feathery in appearance, and consists of minute crystals. Cirrus does not give rain, but its presence is sometimes a sign of an approaching **depression**. In this case, the cirrus is gradually replaced by thicker and lower cloud. In different conditions, when patches of cirrus can be seen in a predominantly blue sky, it is a sign of good and settled weather. Cirrus occurs at heights above 20,000 ft [6,000 m] and sometimes as high as 40,000 ft [12,000 m]. SEE ALSO **cloud**

CIS SEE **Commonwealth of Independent States**

citrus fruit The fruit of any plant belonging to the genus *Citrus*, such as oranges, lemons, limes and grapefruit. Citrus fruits require high temperatures for ripening, and so they are found only in warm climates, such as in Florida, USA, the Mediterranean area or in tropical latitudes.

clay An **argillaceous**, **sedimentary rock** which consists of tiny particles less than 0.0001 in [0.002 mm] in diameter. The particles are very closely packed together, and so it is difficult for water to pass through; for this reason clay is **impermeable**. Clay soils quickly become waterlogged in wet weather. There are many different types of clay, varying in texture, chemical content and origin, such as **alluvium**, **boulder clay** and **loess**.

cliff A high, steep, possibly vertical, rock face, commonly associated with coastal regions, but also occurring inland. The dip of the strata influences the steepness of the slope. If the dip is landward, the slope is likely to be very steep, but it will be more gentle if the cliffs are in hard rocks such as granite, as in Yosemite National Park in California, USA, or **Carboniferous** limestone. Many other rocks form cliffs, as, for example, the chalk "white cliffs of Dover" in Kent, England.

climate The overall weather conditions of a place or region throughout the year.

The daily **weather** records are added together and then averaged out to give a general pattern of the climate. The climate of an equatorial forest is similar to its weather, but in most parts of the world the climate will be different from most of the daily weather. SEE ALSO **continental climate, maritime climate, Mediterranean climate**

climatic change An actual change in **climate** over a period of time for one particular area or the whole world. Historic climatic changes have occurred with the **Ice Age**. In the future, it is thought there will be further climatic change brought on by the effects of **pollution** affecting the **atmosphere**.

clinometer An instrument for measuring inclination or slope.

clint A flat block of limestone between **grikes**. Clints are part of a **limestone pavement** and can be found in many areas where there are outcrops of **Carboniferous** limestone.

closed system An isolated system into and out of which no energy or material may pass. SEE ALSO **open system**

cloud A visible mass of water droplets or particles of ice suspended in the air above the Earth's surface. Clouds form when damp, usually rising, air is cooled; for example, when a wind rises to cross hills or mountains. The types of cloud are classified according to altitude as high, middle or low. The high ones are **cirrus**, **cirrostratus** and **cirrocumulus**. The middle clouds are **altostratus** and **altocumulus**, while the low clouds include **nimbostratus**, **cumulus** and **stratus**. **Cumulonimbus** can be tall enough to occupy middle as well as low altitudes.

coal A **sedimentary rock** formed mainly in the **Carboniferous** period, when dead tree ferns were buried beneath sediment in **anaerobic** conditions which slowed down the processes of decay. The gradual changes which occurred produced coal

►**climate** The world can be divided in five climate zones, and each of these can be subdivided. This division was first made by Vladimir Köppen. Tropical rainy climates (rain forest, monsoon areas and savanna) have a mean monthly temperature above 64°F (18°C). Dry climates (steppe and desert) and have low rainfall and a wide range of temperatures. Warm temperate rainy climates (dry winter, dry summer, or no dry season) have a mean temperature below 64°F (18°C) but above 28°F (–2°C) and the warmest month is over 50°F (10°C). In cold temperate climates (dry winter or no dry season) the mean temperature of the coldest month is below 37°F (3°C) but that of the warmest month is still over 50°F (10°C). Polar climates include tundra, where the mean temperature of the warmest month is below 50°F (10°C), giving permanently frozen subsoil, and the polar climate itself where the mean temperature is below 32°F (0°C), giving permanent ice and snow.

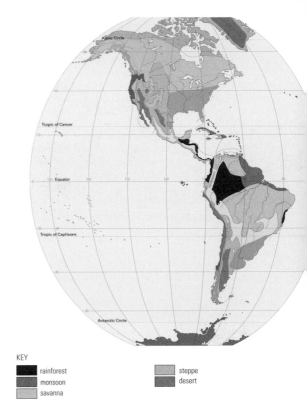

KEY

rainforest	steppe
monsoon	desert
savanna	

from the buried plant material. There are tree ferns growing in certain parts of the world today, such as Malaysia and New Zealand. If these deposits are covered by sediment, it is possible that they may form coal in a few million years. Coal is classified according to its hardness, carbon content, and the amount of smoke given off and the amount of ash left behind when burned. SEE ALSO **anthracite, bituminous coal, brown coal**

coal measures The series of rocks from the **Carboniferous** period which contain seams of coal. The coal measures also contain many strata of clay and sandstone, and coal may account for only 2% of the total thickness.

cocoa SEE **cacao**

coffee Any tree or shrub belonging to the genus *Coffea*, grown for its seeds, which are roasted and ground to produce coffee. The coffee bush requires hot summers – about 75°F [25°C] – for growing and ripening. It is an exhausting crop and requires a rich soil; in the major growing areas of Brazil, many of the soils are from old weathered lavas. The main growing countries are Brazil and Colombia in South America, and Indonesia in Southeast Asia. Western European countries and North America are the chief importers.

cohort SEE **age-sex pyramid**

coking coal A high-quality **bituminous coal** which is converted into coke for use in blast furnaces in steel works.

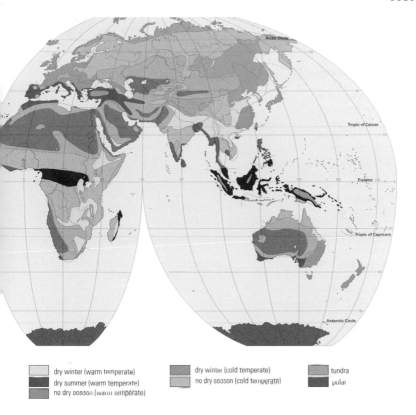

▢ dry winter (warm temperate)	▢ dry winter (cold temperate)	▢ tundra
▢ dry summer (warm temperate)	▢ no dry season (cold temperate)	▢ polar
▢ no dry season (warm temperate)		

col I A **gap** or pass in a line of hills. **2** A calm area between two **depressions** and two **anticyclones**. It lasts only a few hours, as the depressions move in and push it away. It will cause sunny weather during the day, but fog or frosts at night.

cold front The dividing line between two **air masses,** where a cold mass is advancing against and beneath a warmer mass. The upward movement of the warm air is accompanied by condensation, cloud formation and rainfall. In general, as a cold front passes, the wind veers from southwesterly to westerly or northwesterly in the northern hemisphere, and backs from northwesterly to southwesterly in the southern.

collective farming A system of farming found in the former Soviet Union and other communist countries, in which formerly separate smallholdings are worked as one farm, often thousands of acres in size. Each collective has a manager, who is responsible for organizing the work and purchasing machinery.

colony A settlement created by a group of people from a different country. For example, Britain established colonies in Africa (in the late 19th century), Australia, New Zealand and the USA. The French colonized parts of West Africa, and there were Portuguese colonies in southern Africa. Some of the colonies were created in uninhabited areas, but in other cases the colonists found that there were already native inhabitants in the areas of settlement, such as Aborigines in Australia, Maoris in New Zealand and Indians in

▲ **cloud** Stratus (1) is a low-level cloud, usually featureless and grey. Cumulus (2) clouds are small and "fleecy". Cumulus may expand into stratocumulus (3) or giant cumulonimbus (4), commonly associated with heavy rain or hail. At middle altitudes, stratus and cumulus are termed altostratus (5) and alto cumulus (6). Cirrus (7) are named from the Latin for "tuft of hair", which they resemble. Cirrocumulus (8) is often present with cirrus in small amounts. The presence of cirrostratus (9) is given away by haloes around the Sun or Moon.

North America. The colonial influence was not always beneficial; there were wars and bloodshed, and economic wealth was often taken from the colonies and shipped back to Europe. Administrators were not always helpful toward the native population. However, there were some good effects of colonization; for example, roads, schools and hospitals were built, and stable governments were usually created.

combe A small armchair-shaped hollow found on a steep slope, generally an **escarpment**, in a chalk hillside. They are formed as a result of the opening up and enlarging of a **joint** by **weathering**. The weathered material slips to the bottom of the hill, where it may accumulate to form a mound of depositional material called "combe rock." Much of the weathering is the result of **solifluxion**. ALSO CALLED **coomb, coombe**

commercial agriculture The type of farming in which crops or animals are grown or reared for sale in order to earn money. Some commercial farms are quite small, but others are very large, highly mechanized and **capital intensive**. SEE ALSO **subsistence farming, agribusiness**

Common Agricultural Policy (CAP) The rules, regulations and guidelines which help farmers in the member countries of the European Union. They are designed to enable farmers to earn a reasonable income and to encourage them to produce large quantities of essential foods. In several cases, however, the CAP has led to overproduction and consequent surpluses of some products, which must then to be stored at huge expense. The main surpluses have been of wheat, milk, beef, butter, olives and wine.

common land British term for public land which is normally used by commoners for grazing. Historically, common land was not owned by any one person or family, but was available to members of the community. This is still true today of some village greens and riverside meadows, and in upland regions there are often areas of hillside which provide common grazing.

common market A group of states that join to form a single trading market to achieve better economic results for its members than if they were working as individuals. Within the group, restrictions of movement are lessened and individuals, capital, goods and services should move freely. SEE ALSO **European Community, European Union**

Commonwealth of Independent States (CIS)
A replacement in part for the former USSR (Union of Soviet Socialist Republics), set up in December 1991 by Russia, Ukraine and Belarus, and later joined by Moldova, Armenia, Azerbaijan, Kazakhstan, Uzbekistan, Turkmenistan, Tajikistan, Kyrgyzstan, and finally Georgia. Its declaration covered national borders, human rights, ethnic minorities and inter-national law. Committees were also set up to deal with foreign affairs, defence, finance, transportation and security.

communism An economic system historically based on the common ownership of all property. In 1848, Karl Marx developed the idea of communism further: that people should aim for a classless society regulated by the maxim, "From each according to his ability and to each according to his needs." Different types of communism, all based on this one principle, have been tried in many countries.

commuter A worker who travels daily from home to his or her place of work in a town or city. Efficient train services and good

COAL

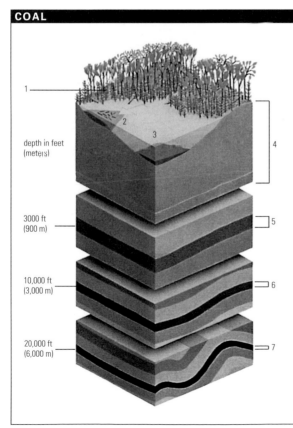

1

depth in feet (meters)

2

3

4

3000 ft (900 m)

5

10,000 ft (3,000 m)

6

20,000 ft (6,000 m)

7

The process of making coal begins with plant debris (1). Dead vegetation lies in a swampy environment and forms peat (4), the first stage of coal formation Underwater bacteria remove some oxygen, nitrogen and hydrogen from the organic material. Debris carried elsewhere and deposited by water forms a product called cannel coal (2). Algal material collected underwater forms boghead coal (3). If the dead organic material is buried by sediment, the weight on top of the peat and the higher temperature will turn the peat into lignite (5). With more heat and pressure at increasing depths, lignite becomes bituminous coal (6) and then anthracite (7).

COMPASS

A compass needle points to the north, If the compass is aligned with the needle, the directions on the compass dial will be true. In fact, the "magnetic north" shown on a compass usually does not correspond exactly to "true north", which is the point around which the Earth rotates at the North Pole. This is because the magnetic and true north poles are not in exactly the sample place.

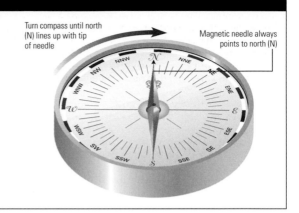

Turn compass until north (N) lines up with tip of needle

Magnetic needle always points to north (N)

road networks are vital to enable large numbers of people to travel to and from work. For example, over 1 million people commute to London, England, each day. A large proportion have journeys of from 30 miles to 40 miles [48 km to 65 km], but some travel far greater distances. The area around a city in which many commuters live is generally affluent and includes the outer suburbs and neighboring towns and villages.

compaction The compression of soil, as, for example, by walkers' feet along a popular footpath. When compaction occurs, it is difficult for rainwater to percolate through the soil, and so muddy patches and puddles develop. Compaction can also occur on farmland if heavy machinery goes over soft or wet ground.

compass An instrument that is used to find direction. A magnetic compass has a magnetized needle which is affected by the local magnetic force (generated by the Earth's magnetic field), and thereby shows its position on a dial in relation to the north and south magnetic poles. A nonmagnetic compass simply shows true north and south, based on the actual geographical position.

component A constituent part of a product; one of the parts which have to be put together in order to make a finished product.

concave slope A slope which is much steeper at the top than it is at the bottom. On a **contour** map, the contours are positioned much closer together at the top than at the bottom. When standing at the top or bottom of a concave slope, it is possible to see the entire slope. Concave slopes are quite common on limestone hills. SEE ALSO **convex slope**

concealed coal field A coal field which is underground. The coal seams lie beneath the strata of other rock types and have to be extracted by shaft mining. SEE ALSO **exposed coal field**

concentric ring theory SEE **Burgess model**

concordant coast A coastline that runs parallel to the hills and structural features immediately inland. When submergence occurs, the sea floods the former valleys and the peaks become chains of islands. Examples of a concordant coast are to be found along the south coast of Ireland, near Cork, in the Croatian province of Dalmatia, and along the Pacific coasts of North and South America. ALSO CALLED **Dalmatian coast**, **longitudinal coast**, **Pacific coast**

condensation The process in which a gas or vapor changes to a denser form as a result of cooling. Mist, fog and cloud, for instance, are formed by the condensation of water vapor in the air. Dew is the result of water vapor condensing on grass or on the ground. Condensation forms tiny droplets of water, but they can merge and grow bigger and heavier; they may then fall as rain.

condensation nuclei Hygroscopic (or water-absorbing) particles, such as dust or salt grains, which may attract water vapor and cause condensation before dew point is reached, thus producing mist, fog or cloud.

confluence The point where two or more streams converge and unite, thus forming a larger, and commonly faster-flowing, stream.

conglomerate A **sedimentary rock** consisting of large fragments which have been cemented together. The fragments are smoothed or rounded, and they are often mainly siliceous in content. Conglomerates form on beaches, where the smooth pebbles are an ideal source of material. SEE ALSO **breccia**

coniferous forest A forest which consists of cone-bearing, usually evergreen, trees with needles, such as pines (*Pinus* spp.), spruces (*Picea* spp.) and firs (*Abies* spp.). There may also be larches which, although coniferous, shed their needles and grow new ones each spring, and there are usually some silver birches in areas of coniferous woodland. Because conifers can withstand cold temperatures, coniferous woodlands are found further north than deciduous woodlands. The largest expanse is in northern Russia and Siberia, and further west in Finland, Sweden and Norway. Smaller areas occur on many mountainsides, for example, in the Himalayas, Alps and Rockies. The major source of softwood for pulp and paper manufacture, conifers also provide timber for building purposes. SEE ALSO **deciduous forest, evergreen trees, taiga**

consequent stream A stream that flows in the direction of the original slope of the land; that is, it follows the dip of the rock strata. Several consequent streams may develop more or less parallel to one another, each flowing down the same major slope. SEE ALSO **subsequent stream**

conservation The careful and thoughtful use of resources. It is important to preserve and manage the landscape and the scenery of the natural **environment**, as well as the soil, minerals and wildlife. Planning and organization are necessary in order to make the best use of the countryside. The conservation of minerals involves using them more slowly, trying to reduce or eliminate waste, and recycling anything which can be reused.

constructive margin SEE **plate tectonics**

consumer industry An industry producing ready goods for the consumer, such as food, clothing and domestic appliances, as opposed to goods which will be used in further manufacturing processes. Many consumer industries are located on industrial parks. They use road transportation and electricity, and this helps to make the location very flexible.

container port A port for handling containers. Heavy lifting equipment and large areas of flat land are required for container ports. Container ships can be loaded or unloaded quickly because their entire cargo is boxed in containers, which are easily lifted straight on to (or off) trains and lorries. Examples of North American container ports are Montreal and Vancouver in Canada, and Long Beach and Los Angeles, California, in the USA.

continental climate The type of climate found in the interior of large continents, that is, hot summers with convectional rainfall and very cold, dry winters, with only light falls of snow. Summer temperatures are about 70°F [20°C], and winter temperatures are 15°F to –5°F [–10°C to –20°C] in the coldest month. Total annual precipitation is about 20 in

▶ **continental drift**
About 200 million years ago, the original Pangaea land mass began to split into two continental groups, Laurasia and Gondwanaland. Over time, these two groups split further over time to produce the present-day configuration of continents.

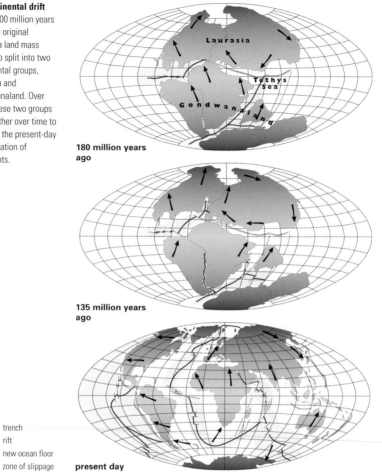

180 million years ago

135 million years ago

trench
rift
new ocean floor
zone of slippage

present day

[500 mm]. The natural vegetation consists of grassland. Areas of continental climate are found in Poland and Hungary, and on the steppes of Russia and the prairies of North America. In the southern hemisphere there are no areas of real continental climate because the land masses are relatively narrow, so that oceanic influences are present throughout. SEE ALSO **maritime climate, Mediterranean climate**

continental drift The slow movement of the continents on the surface of the Earth. The movement, or **drift**, occurs as a direct result of the movement of the underlying plates. Continental drift was first written about by Alfred Wegener in Germany in 1912. He noticed that the western parts of Africa, near the Gulf of Guinea, would fit alongside the coast of northeast Brazil, and discovered that there were geological similarities between the two areas. For a long time his ideas were discredited, but in the 1960s they were shown to be correct. It is now known that there used to be a vast ancient continent, which is referred to as

"Gondwanaland." About 100 million years ago, Gondwanaland fractured, and parts of it began to drift away to form the modern continents of Antarctica, Africa, Australia and South America. There has also been continental drift in the northern hemisphere, North America and Europe formerly having been joined together. SEE ALSO **Pangea**

continental shelf A shallow area of sea which is a gentle continuation of the neighbouring land. The depth of a shelf is less than 1,000 ft [330 m]; beyond that the water becomes much deeper at the continental slope, which descends to the deep ocean bed. A continental shelf may be quite narrow, as, for example, off the west coast of South America, but in places can be more than 100 miles [160 km] wide – for example, in the North Sea and around Britain. The shallow waters contain rich food for fish, and therefore some of the world's major fishing grounds are on continental shelves, including those off the west coasts of Canada and the USA, and of Britain and Japan. Continental shelf areas, including the North Sea and the Gulf of Mexico, have been exploited for oil and natural gas deposits. Unfortunately, there are serious pollution problems in several of the shallow shelf areas of sea, notably the North Sea and the Mediterranean.

contour A line on a map which joins all places of equal height. Maps may use different intervals between contours, depending on the needs of the map. Colored shading can be drawn onto maps in the spaces between neighboring contours; this technique is often used in atlas maps.

contour plowing The practice of plowing upland at right angles to the direction of the slope. Farmers use this method in order to conserve the soil, as

▼ **continental shelf** The regions immediately off the land masses are the continental shelves. There are several different sorts. Some have deep gorges (A); they can be caused by river erosion before the land was submerged by the sea, or by ocean currents. Mud- and sediment-laden water often pours out of major estuaries scouring out gorges. Other regions have shelves with a gentle relief (B), often with sandy ridges and barriers. In high latitudes, floating ice wears the shelf smooth (C) and in clear tropical seas a smooth shelf may be rimmed with a coral barrier like the Great Barrier Reef off eastern Australia (D). Other types of continental shelf often start with a gradual slope, but then drop suddenly. This can be caused by strong offshore currents washing sediment away (E). Continental shelves can also be affected by faulting (F), which disrupts the original shelf.

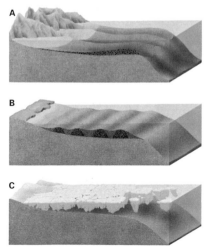

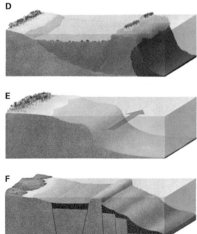

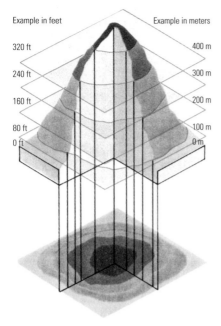

Example in feet   Example in meters

320 ft   400 m

240 ft   300 m

160 ft   200 m

80 ft   100 m
0 ft   0 m

▲ **contour** Recording a three-dimensional shape on a flat surface is a problem that can be solved by the convention of contour scaling. This is shown when the cross-sections of a hill are projected onto a map of the hill. The hill can be envisaged fairly well from such a map, although the "coarseness" of the contour intervals loses some finer detail. The steepness of the sides can be judged by the contour lines on the map.

horizontal furrows prevent or slow down runoff after rainfall. Sometimes different crops may be planted at different levels of the slope; in this way part of the land will still be covered by vegetation when some of the crops have been harvested. Contour plowing helps to reduce **sheet erosion** as well as gullying. It is commonly practiced in parts of the USA, such as Tennessee and Kentucky.

conurbation I A large urban area which has developed as a result of the expansion and merging of two or more urban areas. Urban sprawl can produce a continuous built-up area spreading for several miles. An example in the USA is the Baltimore-Washington metropolitan area. **2** A smaller, but continuous, built-up area.

convection rainfall Rainfall caused by convection currents. As the air rises, it is cooled. When dew point is reached, the water vapor condenses to form **cloud**. If the convection current is strong, a very high cloud may form, with great quantities of water and dust particles. The tall banks of cloud, **cumulonimbus**, can look very threatening; they will produce very heavy showers of rain, often more than 1 in [25 mm] in the space of an hour or less. In tropical areas, most of the rainfall is convectional; but in temperate latitudes convection rainfall occurs only occasionally, in the summer after very hot anticyclonic spells of weather, and will be accompanied by **thunder** and **lightning**.

convenience goods Items which are quite cheap and are required frequently, possibly every day, such as bread and newspapers. ALSO CALLED **low-order goods**

convex slope A slope which is steeper at the bottom than at the top. Standing at the top of a convex hill, it is impossible to see the bottom of the hill because there is always a large blind spot or dead ground. Convex slopes commonly occur on chalk hills. SEE ALSO **concave slope**

coomb (coombe) SEE **combe**

cooperative An organization in which several small farmers pool their resources, equipment and marketing. In this way, they have the purchasing and selling power of a larger organization. A cooperative can help to provide more capital investment as well as a better and more secure income. Cooperatives were first developed in Denmark in the late 19th century, when many small dairy farmers began to work together in groups. Today, they collect the milk from the farms, and either sell it or use it to produce cheese, butter or cream. Some skimmed milk is used for feeding pigs, and the bacon, ham and sausages are also handled by the cooperatives. Other countries now have cooperatives.

coral A tiny marine animal which lives in massive colonies. When coral polyps die they leave a very small deposit of **calcareous** material, which forms limestone rock; this in turn may build up to form a **coral reef**. Corals can be seen as fossils in some **Carboniferous** limestone rocks. This indicates that at the time the rocks were being formed they were on the bed of a tropical sea, since most corals can live only in warm water. Some coral fossils have even been found high on the slopes of Mt. Everest.

coral reef A large ridge of coral found offshore that may be completely covered at high tide and only partially exposed at low tide. The coral forms best in shallow water, where there is good sunlight and temperatures of over 70°F [20°C]. The coral reefs themselves are often very deep; this happens when the coral forms and then sub-

sides, causing new coral to form on top. Layers are therefore built up. Fringing reefs are found close to the shore, while barrier reefs lie further away. Coral sand is formed when coral breaks up into small fragments. SEE ALSO **coral, barrier reef, atoll**

cordillera A mountain chain. This term is particularly used to describe the parallel mountain ranges of the Andes in South America or the series of mountain chains in western North America, including the Rockies, the Sierra Nevada and the Coastal Ranges.

core I The central part of a region. The outer or more isolated parts are called the **periphery**. John Friedmann wrote about the "core-periphery idea" after his studies of economic development in Venezuela. The core is the area where most of the economic development and growth take place. In **Third World** countries, it may be the only area with an **infrastructure** of roads,

▼ **Coral reefs** grow in warm waters over 70°F (20°C), thriving at about 75°F (25°C). Reefs grow beyond the tropics in sufficiently warm currents. Corals do not grow in freshwater discharges and cannot grow in water full of sediment, which smothers the fragile reef organisms. Although the optimum depth for growth is a few feet below the surface, where oxygen and sunlight are abundant, coral can grow at up to around 150ft (40m). In ideal conditions, healthy reefs grow up to 1in (25mm) in a year.

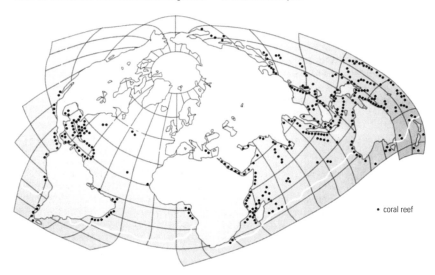

• coral reef

electricity, water supply, etc. In advanced countries, all aspects of the infrastructure will be more widely available, but, even so, there are particularly favored areas where growth is more likely to occur. The core in England would be the southeast, around London. On a larger scale, the core of the European Union would be a zone encompassing southeast England, Belgium, the Netherlands, northeast France and western Germany. It is often difficult to create developments in peripheral regions because there will be transportation problems and possibly labor shortages. Governments often give aid to encourage companies to set up industries in the more isolated places. SEE ALSO **growth poles**. **2** (*in physical geography*) The inner part of the Earth, where temperatures are very high. The rocks are mostly made of iron and nickel.

Coriolis force A force resulting from the rotation of the Earth which deflects moving bodies to the right in the northern hemisphere and to the left in the southern hemisphere. This affects winds and ocean currents.

corn l A widely grown cereal crop, *Zea mays,* bearing large spikes of cobs of grain. Corn is a staple food in many parts of Africa and in some parts of South America. It is grown in vast quantities in the USA, notably in the area of the Midwest known as the "Corn Belt," where it is used primarily as a feed for cattle and pigs. Corn requires a wet growing season, with 20 in to 30 in [500 mm to 750 mm] of rain, and needs summer temperatures of 70°F [20°C] or more in order to mature and ripen. ALSO CALLED **maize 2** (*in Britain*) Any of the principal **cereal** crops, that is, wheat, oats or barley.

corrasion Mechanical erosion caused by loose material, such as sand or pebbles, during transportation. As this material is carried by wind, water or ice, it scrapes against the bed or sides of the river or glacial valley, or against the cliffs and the shore. Corrasion also occurs in deserts, where it is caused by wind-blown sand. The

sand and pebbles carried in the load of a river can cause both vertical and lateral corrasion.

correlation A statistical technique for comparing two or more sets of data. The variables can be compared in a precise and quantitative way. There is generally one dependent variable, the others being independent. The strength of the link or relationship is quantified between +1 and −1. A strong inverse relationship or negative correlation (where one variable increases as the other decreases) would be −1, while +1 would indicate a strong direct relationship or positive correlation (where an increase or decrease in one variable is matched by a similar one in the other); 0 would reveal the absence of an explanatory relationship. SEE ALSO **Spearman rank correlation**

corrie SEE **cirque**

corrosion Chemical erosion leading to the disintegration of rocks, such as by the action of running water or by **solution**. Corrosion is particularly common in limestone.

cottage industry A small-scale industry which is carried on in the home of the worker. Many woollen and cotton goods, as well as baskets and pots, are made in this way. They are generally marketed and sold by a middleman or cooperative organization. Cottage industries used to be common in the Scottish Highlands and in Ireland, and there are still examples to be found. Many continue to flourish in Canada and the USA. India has many small-scale domestic industries, and they can also be seen in many parts of Africa, as the rural people try to earn extra money.

cotton Any bush or tree of the genus *Gossypium* that is grown for the fluffy white fibrous substance which surrounds its seeds. The cotton bolls (seed heads) are still picked by hand in many places since it has proved difficult to develop a machine which will pick only the ripe bolls, leaving

COROLIS FORCE

One example of the Coriolis force occurs because wind will tend to drag ocean surface water after it. However, as soon as the water starts to move, the Coriolis force throws it off at an angle. Each successive thin sheet of water beneath the surface layer begins to move because it is linked by friction to the layer above. But each layer is thrown even further off course by the Coriolis force. So the surface current moves at an angle of 45° to the wind and at about 2–3% of the wind speed. Lower layers of water move more slowly, at an increasingly eccentric angle. The resulting vertical profile of the motion is known as the Ekman spiral. At a depth called the Ekman depth, the water flows in the opposite direction to that of the surface current and at 0.043 times the speed of the surface current.

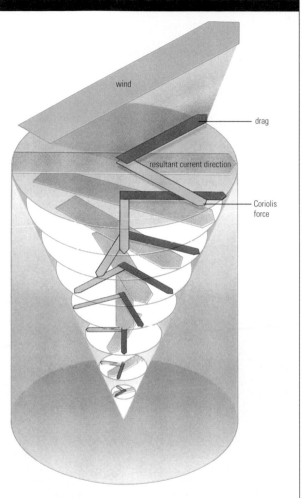

wind

drag

resultant current direction

Coriolis force

the unripe heads to mature. After picking, the cotton is ginned to clean it. Then it has to be combed and spun into thread before it is finally woven into cloth. The major cotton-growing area is the Cotton Belt of the southern USA, comprising 18 states, with Texas producing the most. The subtropical climate of this area means that summer temperatures are high (75°F [25°C]) and the growing season is long, with over 200 frost-free days. There is a normally dry period before harvesting is due, which is useful, as too much rain can damage the boll. Some of the cotton is grown on plantations which used to employ slave labor in the 19th century. There are still large estates in the Cotton Belt, but they also grow other crops, such as corn, soybeans and peanuts, and often raise cattle.

counter-urbanization The movement of people away from cities into smaller towns and villages. Counter-urbanization is a trend among the wealthier sections of the population in North American and Western European countries and is not associated with **Third World** countries, where there is still a strong flow of people into the cities.

cover crop A crop grown to protect the soil from heavy rain, wind and erosion. Often consisting of a quick-growing plant, it may be grown after the main harvest of **cereals**, when large areas of land would be left bare. Sometimes strips of land on a hillside are left with a cover crop, while the remainder of the slope is being harvested. The most useful cover crops are **legumes** because they grow quickly, have good rooting systems for holding the soil and also add nitrogen to the soil.

crag A steep-edged rocky outcrop on a hillside. It will be the result of **erosion** and **weathering**, and is most likely to be found in areas of hard rock. Glaciated areas often have exposed rocks on **truncated spurs**.

crag-and-tail A **crag** with a tail of glacial material on the down-ice side. The debris may have accumulated as a **glacier** or **ice sheet** scraped over the hard lump of rock. Alternatively, **erosion** may have occurred on one side only, the down-ice side having been protected by its situation. An outstanding example is the crag-and-tail on which Edinburgh Castle in Scotland is located.

crater I A depression at the top of a **volcano**. Whether at the summit of a mountain or simply as a hollow in the ground at low level, it will form the top of a tube or vent through which the erupted material passes to the surface. When the volcano ceases to erupt, the vent and the crater may be filled with solidifying rocks. Examples of low-level craters can be seen in the Eifel district of Germany. **2** A hollow on the Earth's surface caused by a falling meteorite. There are also many craters on the Moon.

crater lake A lake which has formed in a volcanic crater. If the **volcano** is still active,

any eruption will begin with the ejection of large quantities of water out of the crater. Crater Lake in Oregon, USA, is a good example of a crater lake. Other examples can be found in Indonesia, Italy and Germany.

creep The slow downhill movement of **soil** or **regolith**. The movement is caused by gravity, but is helped by water acting as a lubricant. The rate of movement is very slow, measured only in fractions of an inch per month, or even per year. The soil which has slid downhill may accumulate against walls or other barriers and at the bottom of steep slopes. SEE ALSO **soil creep**

crevasse A vertical crack or fissure in ice. It may be many feet in depth and will continually change as the ice moves downhill. Crevasses form when the slope of the bedrock over which the ice is moving becomes steeper. If the gradient lessens, then the crevasses tend to close up. Crevasses caused by a change of slope normally run across the valley and are transverse. Longitudinal crevasses form when the valley broadens out and the ice stretches to fill up the extra space. As crevasses move downhill, they change angle slightly, and so it is possible to have crevasses running in any direction. Crevasses are largest on mobile glaciers and are potentially very dangerous for anyone crossing a **glacier** or **ice sheet**. SEE ALSO **icefall**

croft A small farm in northwest Scotland, on the Scottish islands or in western Ireland. It consists usually of a house with a small plot of land, where potatoes, hay and oats may be grown, and there will be grazing land for sheep and possibly a few cattle. The crofters will grow most of their own food, and possibly make their own clothing, from wool. They will probably get additional food from fishing. Nowadays, many crofters are involved in the tourist industry, and so they can earn some money; but 50 years ago they were wholly dependent on **subsistence farming**.

crop rotation A system of farming in which each field is used for a different crop or animal every year. There may be a planned cycle of planting over a period of two to eight years. The first well-known method of crop rotation was developed in Norfolk, England, in the 18th century. The four-year cycle was a **root crop** – generally turnips – followed by barley, then grass and clover, and finally wheat. The rotation gave two **cereals** and a good fodder crop, as well as grass which could be used as pasture, and clover, which would restore nitrogen to the soil. It was a great improvement on previous systems, which had involved leaving the land **fallow** to allow it to rest and recover. It meant all land could be used in a productive fashion, while maintaining the quality of the soil. The main disadvantages of crop rotation are that farmers are involved in a large range of activities and therefore cannot afford specialized equipment, and that as they have many small fields they cannot use large-scale methods. On the other hand, the advantages are that the quality of the soil is maintained and that animal manures are available, which will reduce the dependence on chemical fertilizers. Weeds, pests and diseases are more easily kept under control, again reducing the need for chemical controls.

cross-section A technique used in map work to show relief in a clear and simple fashion. Cross-sections taken from maps may have a horizontal scale equivalent to that of the map, such as 1:50,000 or 1:25,000. The vertical scale is exaggerated in order to show up the undulations. However, the exaggeration must not be too great, otherwise quite small hills will look like large mountains.

cross-wind A wind that blows at almost exact right angles to the course of a moving object; for example, across the flight path of an airplane.

crude birth rate SEE **birth rate**

crude death rate SEE **death rate**

crust SEE **lithosphere**

cuesta A hill with a steep slope ("scarp") on one side, and a more gentle slope ("dip slope") on the other. A cuesta has an asymmetric cross-section. It may have been formed by **folding**, but is more likely to be the result of tilting, followed by **erosion**, different rocks having eroded at different rates to form the steep scarp on the exposed edge of a resistant stratum of rock.

cultural geography The study of the effect of the human impact on the natural landscape; for example, the domestication of animals, farming and the building of towns and cities.

cultural landscape The result of the effect of human activity on the natural landscape. This builds up over time, with one generation/society affecting the landscape for the next. Very few areas of the world exist now that are unaffected by human activities.

cumulative causation The theory that an initial advantage can produce a sequence of developments, which in turn lead to further advantages and developments. Cumulative development of this kind occurs in many industrial regions. For example, the coal which gives rise to a steel industry may also support a chemical factory, which would also help steel-making. Then engineering works may be set up to provide equipment for the steel and chemical industries. Meanwhile, electricity supplies will have been connected, and schools, hospitals, roads and railroads will have been built. Anyone then wishing to start a new enterprise would choose to go to this area because of the amenities and advantages it has to offer. **Gunnar Myrdal**, who formulated the theory of cumulative causation, devised a development model following this principle. In **Third World** countries, most developments occur near the major city or cities, which are the only locations with **infrastructures**. The accumulation of development in the favored areas leads to great regional inequalities in all countries. An attempt to develop new areas requires new **growth poles**.

cumulative frequency curve A frequency curve produced by adding the value of each successive group to that of the last. For example, if 15% of the inhabitants of an area were aged under 15, another 15% were aged 15–25, and a further 10% aged 25–30, etc., the points plotted on a graph would be 15, 30 (15 + 15), and 40 (15 + 15 + 10).

cumulonimbus A dense cloud with a flat base and a high, fluffy outline, commonly associated with thunderstorms. SEE ALSO **cloud, convection rainfall, cumulus**

cumulostratus SEE **stratocumulus**

cumulus A dome-shaped, flat-bottomed cloud of considerable vertical growth that usually occurs at low altitudes. Small white cumulus are commonly associated with fair, dry weather, but large towering cumulus may develop into **cumulonimbus**, which bring heavy rainstorms. SEE ALSO **cloud**

current A quantity of water or air moving in a specific direction. On a global scale there are **ocean currents**, while on a smaller and more local scale there are currents in rivers caused by the horizontal flow of water in particular channels.

cut-off SEE **ox-bow lake**

cwm SEE **cirque**

cycle of erosion The continuous process by which landscapes are formed and then gradually transformed by erosion. Mountains are formed by uplift, folding, faulting and volcanic activity, all of which are related to **plate tectonics**. Once the high land has been formed, rivers, ice and all the other agents of **erosion** and **weathering** begin to denude the landscape. When it is still high, but dissected by steep **V-shaped valleys**, it is described as a young or youthful landscape. Various methods of transportation remove the eroded materials and expose more rocks for denudation. Gradually, the high land will become lower and the valleys will widen to form a mature or middle-aged landscape. Eventually, after millions of years, most of the high ground

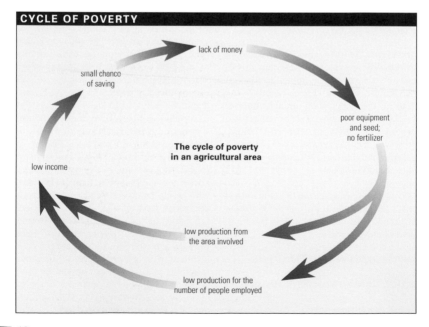

CYCLE OF POVERTY

lack of money

small chance of saving

poor equipment and seed; no fertilizer

low income

The cycle of poverty in an agricultural area

low production from the area involved

low production for the number of people employed

will have been eroded and the river valleys will be broad plains. This old-age landscape, with mostly flat and low ground but just a few isolated residual hills, is called a **peneplain**. The sequence of events described is a very simplified and theoretical view of landscape evolution, because uplift and erosion take place simultaneously.

cycle of poverty A chain of circumstances in which poor people are unable to grow much food because of a lack of money for seeds and equipment, and therefore cannot earn money from their low production or do more farming in order to produce the food they need to sell and eat. They struggle for survival from one year to the next, are trapped in a vicious circle and cannot escape without help from outside. This cycle may be passed on from one generation to the next.

cyclone A system of atmospheric low pressure which occurs when a cold **air mass** moving southward from the Arctic or northward from the Antarctic meets a warm air mass moving from the tropics to form a circulating air mass characterized by relatively low pressure at the center, and by counterclockwise wind movement in the northern hemisphere and by clockwise motion in the southern hemisphere. There are two types of cyclone: the **depression**, associated with temperate latitudes, and the **tropical cyclone**, which is much more violent but usually affects a smaller area.

D

dairy farming A system of farming in which cattle are kept for their milk, which may also be used for butter, cheese, cream and yogurt. It is normally **intensive** and dairy farms are relatively small. Rich grass is needed to feed the cattle and mild winters are an advantage, so that the cattle can stay outside all year. In some mountainous areas, such as the Alps, the cattle are kept indoors for up to six months. For this reason, many farmers try to make as much **silage** and hay as possible during the summer months, to give to the cattle in winter.

dale A valley in northern England or southern Scotland.

Dalmatian coast SEE **concordant coast**

dam A barrage, especially one associated with the generation of **hydroelectric power**. Some dams may be enormous (for example, Aswan in Egypt) and create lakes over 30 miles [50 km] long, whereas others may be no more than a few meters in length and form a lake which dries up for half the year.

Date Line SEE **International Date Line**

death rate The number of deaths in a year for every 1,000 head of the population. In poorer countries the figure is often higher than in richer countries. For example, in 1997, the US figure was 9, but in Malawi the death rate was 25. ALSO CALLED **crude death rate**

debris Accumulated material, such as scree, gravel and sand, formed by the break up or disintegration of minerals and rocks. This material is then often transported (by air, water or ice) from its place of origin and redeposited.

decentralization A movement away from a (usually **urban**) concentration. This may be a planned policy by governments to stimulate growth in depressed regions away from the main cities; they may offer incentives, such as loans or tax relief, to encourage businesses to relocate. Decentralization may also occur when people no longer wish to live in congested urban areas, where there may be a high level of noise, pollution and crime.

deciduous forest A woodland consisting of trees which shed their leaves. Most of the trees associated with deciduous woodlands are broad-leaved, such as oak, ash, beech, etc., and in **temperate** latitudes they shed their leaves annually in the fall because the winters are too cold and frosty for growth to continue. However, in some **monsoon** lands, such as India and Myanmar, there are trees which shed their leaves in the hot season in order to avoid the excessive loss of moisture caused by **transpiration**. Near the Equator, because of the constant climatic conditions, the deciduous trees lose their leaves sequentially rather than in one season, since there is heat and moisture all year around. Although the trees shed some leaves each month, they are able to grow new ones, too, as if the trees were going through all four seasons at the same time. SEE ALSO **coniferous forest, evergreen trees, taiga**

deep A long and fairly narrow depression in the ocean floor, of over 18,000 ft [5,500 m] in depth. Deeps (also known as "ocean trenches") occur particularly where there has been plate movement. An example is the Mariana Trench, near Guam in the Pacific Ocean, which has a depth of 36,161 ft [11,022 m]. SEE ALSO **ocean**

deflation The removal of tiny particles of dust or sand by the wind. It is a common process in deserts; in some places, deflation removes all the depositional material, leaving a surface of bare rock. Deflation hollows can be formed in areas of **sand dunes** in coastal regions when the wind transports large quantities of sand.

DELTA

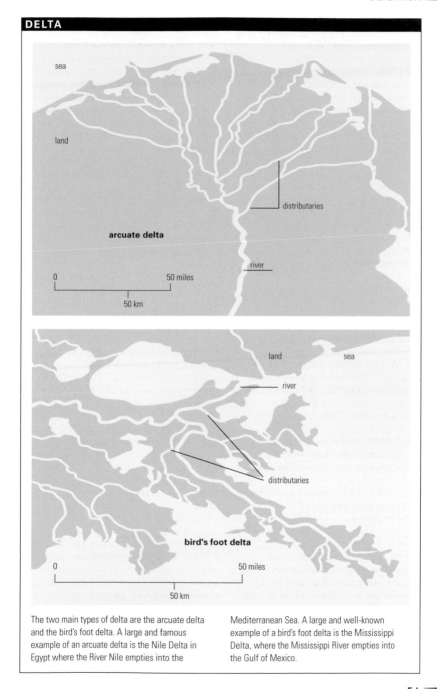

The two main types of delta are the arcuate delta and the bird's foot delta. A large and famous example of an arcuate delta is the Nile Delta in Egypt where the River Nile empties into the Mediterranean Sea. A large and well-known example of a bird's foot delta is the Mississippi Delta, where the Mississippi River empties into the Gulf of Mexico.

deforestation The cutting down and removal of trees, usually to clear the land for some other use, such as agriculture, urban construction, industry or road building. Much of Britain was cleared for agriculture during the Middle Ages. Nowadays, deforestation is occurring mostly in **Third World** countries as new land is opened up for agriculture. In many places, deforestation has exposed the soil to heavy rain, which has led to severe **erosion** – as, for example, in many parts of southern Africa, northeast Brazil and the Amazon basin in South America. Where steep slopes have been deforested, **soil erosion** has become a problem; this has happened in many countries, including Italy and New Zealand. In some fairly dry regions, the soil erosion has been particularly serious, since **desertification** has followed; for example, in the **Sahel** in central Africa. Another result of deforestation is increased **flooding**. Trees and their roots slow down the rate of runoff and **throughflow**, but in places which have lost their trees, runoff is much faster and floods are more extensive. This is true in many parts of Southeast Asia and also in Amazonas province in Brazil. The removal of forests on Himalayan slopes in Nepal and northern India has caused damaging floods in the Ganges valley in India and over much of Bangladesh.

degradation 1 The lowering of a surface by **weathering** and **erosion**. **2** In soil, a deterioration as a result of **leaching**.

delayed runoff The proportion of rainfall which sinks into the ground and gradually percolates through the rocks and back into a stream river. SEE ALSO **surface runoff**

delta An accumulation of silt deposited on the seabed at the mouth of a river. When a river reaches the sea, its speed is reduced and so it has to deposit some of its load. If this deposition builds up above sea level, it will form mud banks, which may grow into a delta. Deltaic deposition will fill in an estuary and gradually spread seaward. Vegetation will colonize the depositional area and convert the sediments into firm ground on which farming can take place. Deltas form most easily in shallow seas, and also where the tides and currents are not very strong, otherwise the depositional material may all be carried away. Very large deltas, covering hundreds of square miles, are found along some coasts – as, for example, those of the Mississippi, Ganges, and Nile. Deltas which are triangular in shape are known as "arcuate deltas," such as the Nile delta, but others, including the Mississippi, are called "bird's-foot deltas" because they create narrow mounds of deposition protruding out to sea. Many deltas become important for farming as they have been created by alluvial deposits, which are generally a source of rich and fertile soils. The Ganges delta in Bangladesh is very productive, though often affected by **flooding**.

demersal fish Fish which live near the bottom of the sea, such as cod, flounder, halibut, plaice, sole and whiting; they are caught by trawling.

demography The geographical study of human populations by mathematical and statistical techniques, to find their size, composition and distribution.

demographic transition The change in the pattern of population growth. In agricultural societies in a preindustrial era, there are high **birth rates** and high **death rates**, so the population remains fairly constant. Britain was at this stage before the **Industrial Revolution**, and some parts of Africa have not yet advanced beyond it. With improvements in education, diet, medicine and some technology, the high death rate begins to fall rapidly, but the birth rate remains high. This means that there is a rapid increase in population. Many African, Asian and Latin American countries have just gone through this stage of development. In the more advanced countries, education continues to improve and a high proportion of the population now live in the **urban** areas. Standards of living are higher, there is more employment and income, and there is no longer the

DEMOGRAPHIC TRANSITION

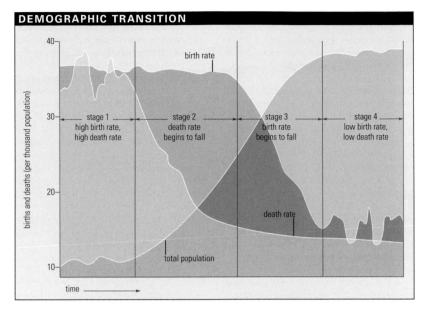

necessity to have large numbers of children. The birth rate falls and gradually becomes closer to the death rate. Eventually, the birth rate will be comparable to the death rate and the population becomes stable. SEE ALSO **zero population growth**

dendritic drainage A **drainage** pattern which looks like a tree when viewed from above. The main river is like the trunk, the major tributaries represent the branches, and the minor tributaries are like the twigs. Dendritic drainage patterns are most likely to develop in areas where the rock types are similar so that the rate of **erosion** is fairly uniform. An example of dendritic drainage is the Mississippi River in the USA.

denudation The wearing away of land by **weathering** and **erosion**. Denudation is a broad term and includes all the natural agencies, such as sun, rain, wind, rivers, frost, ice and sea, as well as heating and cooling, freezing and thawing, **solution**, **abrasion**, **corrasion** and **corrosion**. In addition to weathering and erosion, the removal of material, that is, the transportation, is also part of denudation. Together with **deposi-**

tion, denudation is the major process which creates the Earth's landscape.

départements The regional government divisions into which France is split. They were drawn up in 1790 as areas in which everyone could reach the central town within one day.

depletion Used in geography to describe the reduction in numbers of resources; for example, reserves of **fossil fuels**, which are being used up fairly rapidly as energy use increases.

depopulation An absolute reduction in the number of people living in an area.

deposition The laying down of material which has been removed by **denudation**. Most of the material will be sediment and therefore a possible source of **sedimentary rocks** when consolidated. After denudation has taken place, the material is transported, sometimes over quite considerable distances, before being deposited. Most of the material will eventually be dumped on the seabed by rivers. However, there will be some deposition on land – for example, **silt** on flood plains, **till** deposited by ice, **loess** deposited by wind. Together with denudation, deposition is the major process which creates the Earth's landscape.

depressed region A formerly thriving and successful industrial area which has declined as a result of one or more of its industries becoming uneconomic. Several of the old coal and steel regions of the USA are examples of depressed regions, where the decline of industry has led to high rates of unemployment. SEE ALSO **development area**

depression An area of low **atmospheric pressure** in **temperate** latitudes. Depressions are formed when two contrasting **air masses** come into contact with one another to form a circulating mass of air. In the center the air rises, and the atmospheric pressure is lowest (usually 970 mb to 980 mb) where the air is rising most strongly. The two air masses which form the depression will be associated with two fronts – a **warm front** and a **cold front**. The warm front is the leading front and gives a belt of **cloud**, and possibly **rain**. The temperature rises at the front and the wind direction changes from southeasterly to southwesterly. These marked changes of wind direction and temperature are called **discontinuities**. Behind the warm front is the **warm sector**, where a low layer of **stratus** cloud will give gray and gloomy weather, but probably not any rain. This is followed by the cold front, with tall **cumulonimbus** clouds and the likelihood of heavy rain. Behind the cold front will be brighter, showery weather, but colder and accompanied by northwesterly winds. Gradually the cold front catches up with the warm front to form an **occlusion**. Depressions move at 20 mph to 30 mph [30 km/h to 50 km/h] and generally travel from west to east or southwest to northeast. They occur in all the major oceans in **temperate** latitudes and generally cover an area of 300 to 1,200 sq miles [800 to 3,000 sq km]. Deep depressions with lower atmospheric pressure can give rise to winds

of 50 mph to 100 mph [80 km/h to 160 km/h], but shallow depressions are associated with winds of 20 mph to 30 mph [30 km/h to 50 km/h]. SEE ALSO **cyclone**

desalination The process of removing salt from sea water. In a number of areas, where there is a shortage of fresh water, salt is extracted from salt water by means of desalination plants. They also produce piles of salt in addition to the fresh water. Desalination plants can be found on islands such as Lanzarote in the Canary Islands, and in some desert countries such as Kuwait and Saudi Arabia.

desert An area with a dry climate, which is sometimes defined as having a total annual rainfall of 10 in [250 mm] or less. The effectiveness of 10 in [250 mm] can vary, however, depending on whether it falls within a short time or over a prolonged period. A desert is almost barren, although there are very few places with absolutely no vegetation – this occurs only on some patches of moving **sand dunes** or on bare rocky areas. Most deserts contain tufts of grass and scattered, usually thorny, bushes which have the ability to withstand long dry spells. In some deserts there are cactus plants and various species of flowering plant which have a very short life cycle. They spring up quickly after a shower of rain and go through their complete life cycle in a few days, before dying down and leaving seeds to lie dormant until the next rain. Some deserts contain large areas of sand dunes, but most deserts are rocky. There are bare rock areas, but generally there will be a cover of loose stones with patches of moving sand. The stones and sand are gradually broken down by mechanical weathering and wind action. The world's major deserts, including the Australian, Kalahari and Sahara, are found in the **horse latitudes**, where the permanent high pressure causes drought throughout the year. Deserts occur in various west coast areas where the influence of cool currents offshore makes the land even drier, with less than 4 in [100 mm] rainfall in places; for example, the Atacama Desert in

Peru (with the Humboldt Current offshore) and the Namib Desert in southern Africa (with the Benguela Current offshore). There are deserts in the middle of the largest continents where no onshore winds can reach to bring any rainfall; for example, the Gobi Desert in Asia, and smaller examples in Nevada and New Mexico in the USA. Some cold regions are also regarded as deserts because they have less than 10 in [250 mm] annual precipitation. Antarctica and Greenland are ice deserts, and even the **tundra** areas of northern Canada, northern Siberia and elsewhere have some similarities to deserts, although they have cool summers and very cold winters.

desertification The process by which a desert gradually spreads into neighboring areas of semidesert, transforming them into true desert. The change may result from a natural event, such as the destruction of the vegetation by fire or a slight cli-

matic change, but it occurs most frequently as a result of human activity. In many semi-arid areas and dry grasslands, the vegetation becomes overgrazed by domestic animals, so that the land is left bare. Wind and rain then erode the soil, removing any residual fertility, so that no new vegetation can survive. Once the vegetation and soil have gone, the land becomes desert. An additional problem is the use of trees and bushes for firewood by the local people. They must have wood to make the fires they need for cooking. With an increasing population there is an increasing need for fuel, and so landscapes become depleted of trees. Once it has been destroyed, restoring the land is a long and slow process. Extensive desertification has been taking place in the **Sahel**, the southern edge of the Sahara Desert, and there are also examples in southern Africa, India and many other places where there are too many people trying to make a living from an inadequate landscape. The land needs to be rested and

DESERTIFICATION

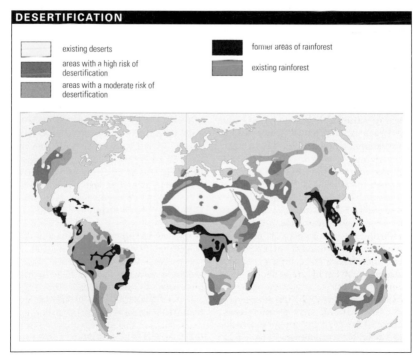

existing deserts

areas with a high risk of desertification

areas with a moderate risk of desertification

former areas of rainforest

existing rainforest

the size of herds restricted, if more land is not to turn into desert.

desert soil Due to a lack of rainfall, desert soils have poorly developed **horizons**, with little vegetation and a thin organic layer, the climate being too dry for chemical **weathering** or the formation of **humus**.

destructive margin SEE **plate tectonics**

destructive wave A wave action which has a destructive effect on the shoreline because the **backwash** is more powerful than the swash. If a series of waves come up a shore in quick succession, they break with an almost vertical plunge, washing shingle and sand downshore, and removing material out into slightly deeper water.

developed countries Those countries with a highly developed industrial sector and a high proportion of the population living in **urban** areas. Agriculture is likely to be commercial and mechanized, with **subsistence farming** constituting a minority activity. **Infrastructure** will be widely spread. Most of North America, Europe, Australasia and Japan could be described as developed, as well as parts of China, Africa and Latin America.

developing countries The poorer **Third World** countries in which numerous economic developments continue to take place. Brazil, Nigeria and many other countries have been developing rapidly, as industrial growth and urban growth have shown some similarities to Western Europe and North America. Loans from the World Bank and foreign countries are often vital for such developments, and several of the developing countries are now heavily in debt.

development area (*in Britain*) A depressed area of high unemployment where the government provided finance and subsidies to encourage the development of new industries. New towns were

created and many trading estates were built. Development areas no longer exist in their original form, but many of them are now called "assisted areas" and can receive government help with a variety of development schemes.

dew Moisture condensed from the air and deposited on grass and other plants, especially during the night. Dew is formed in calm and settled conditions generally associated with anticyclonic weather. During the night, the temperature falls and the air is unable to hold as much water vapor; some condenses on to plants as dew, and some may form **mist** or **fog**. The cooling which causes dew is the result of radiation during the night. The air cools from the ground upward, and if dew point is reached, **condensation** will occur. It is most frequent after sunny days, especially in the fall. Dews are very common, and heavy, in deserts, and there are some plants which survive on the moisture from dew. If the temperature falls below freezing point, the dew would become **hoar frost**.

dew point SEE **absolute humidity**

dew pond British name for a small, usually artificial, pond in which water can accumulate for sheep and cattle. Normally lined with clay or concrete in order to hold water, they are found on the chalklands of southern England and on some limestone hills. In spite of the name, the water is derived mainly from rainfall, not dew.

differential erosion The type of **erosion** which occurs as a result of different types of rock wearing away at different rates. Coastlines of erosion are rarely straight because of differential erosion, and it is a phenomenon which also affects inland landscapes. Hills and undulations are the result of variations in resistance to erosion shown by different types of rock. **Igneous** intrusions are normally quite hard and so **sills** and **dikes** generally form higher ground than their surroundings.

diffusion The dispersion of a phenomenon, or phenomena, through time from a center to other areas. The phenomenon may be a disease, a belief, a language, a technique, etc. The dispersal occurs in a "diffusion wave," as the new factor moves from one area to the next; it will weaken gradually due to the passage of time or over difficult territory or from competition from another diffusion wave.

dike I An **igneous** intrusion which is vertical or near vertical. It is a thin seam of rock which, in molten form, came from a **magma** reservoir and forced its way through the existing rock strata. It is discordant to the structure of the other rocks, and is likely to be harder, so that after long-term **erosion** it will form a narrow ridge or slight undulation. Dikes are found in western Scotland, Iceland and many other present or former volcanic areas. ALSO CALLED **dyke** SEE ALSO **sill** [def. 2] **2** A drainage ditch. **3** An embankment which has been built up to prevent flooding. It may be designed to stop a river from flooding or to keep out the sea, as in coastal areas of the Netherlands.

dip The steepest angle of a tilted stratum of rock. It is measured from the horizontal, 90° being vertical.

dip slope SEE **cuesta**

discharge The water flowing down a river channel. The discharge is expressed as $Q = AV$, where:

Q   is the discharge, generally in cubic feet per second (ft^3/s) or cubic meters per second (m^3s^{-1})
A   is the cross-sectional area of the channel
V   is the mean velocity of the stream

The mean discharge of the River Amazon is 6 million ft^3/s (170,000 m^3s^{-1}).

discordant coastline A coastline that runs at right angles to the structural features of the landscape immediately inland.

It shows up most clearly where there are lines of hills or mountains running inland. If submergence takes place, the valleys will form large inlets, known as **fjords** or **rias**, depending on how they were originally formed. There are several good examples of discordant coasts on the edges of the Atlantic Ocean, for example, in southwest Ireland, Brittany in France, and northwest Spain. ALSO CALLED **Atlantic coast**, **transverse coast**

diseconomy An unfavorable effect usually in monetary terms. "Diseconomies of scale" occur when a business becomes too big and profits may be lessened.

dissected plateau SEE **plateau**

distance decay The pattern of diminishing influence of an urban area in proportion to its distance from a neighboring urban area. This is because of transportation problems and the time taken in traveling from one place to another. The attraction of a zoo, country park, etc., lessens with distance.

distributary A branch which flows from a main river and does not rejoin it, distributing the water and eventually carrying it to the sea. Distributaries may be narrow and shallow, but in some cases are large enough for navigation, as, for example, in the Ganges delta in India.

diurnal range The difference between the maximum temperature during the day and the minimum temperature at night in a 24-hour period.

diversification The broadening out of a business or industry to encompass a greater range of goods or activities. In an industrial area where there has been a concentration on one speciality, it is advisable to develop different industries, in case the main industry suffers from a slump. This happened in many old coal and steel areas and so a variety of industries were set up on industrial parks. In agriculture, **monoculture** can be very successful commercially for a few years, but one pest or disease may destroy the entire crop. It also seems likely that repeatedly growing the same crop is harmful to the soil. By growing a variety of crops (and perhaps introducing a variety of animals) the danger from pests can be reduced, and by rotating crops the quality of the land can be maintained.

doldrums The low-pressure area near the Equator. The overhead sun keeps the equatorial latitudes permanently hot, and so there is always rising air to create low pressure. The rising air causes convection storms, and so rain is frequent, often daily. For this reason, these latitudes are naturally covered with forest. Because of the seasonal changes in the position of the overhead sun, the main center of low pressure moves a few degrees north of the Equator in June and a few degrees to the south in January. The northeast and southeast **trade winds** converge in the doldrum belt, contributing to the mass of rising air. The meeting point of the two trade winds is the "intertropical convergence zone."

Domesday Book A detailed survey (and therefore important historical record) of England – excluding Cumbria, Durham and Northumberland – in two volumes, compiled in 1086–7 on the orders of William the Conqueror, King of England. It includes ownership, extent and value of estates, local customs and a census of householders.

domestication The taming of former wild animals and plants so that they can live with and be used by humans.

dormitory town British name for a settlement a few miles from a large city in which many of the inhabitants are commuters. There will often be a shortage of stores and amenities in the dormitory settlement. Dormitory settlements can be found around most large cities.

dot map A map on which dots representing a specific value are used to show distribution. If showing the distribution of

sheep, for example, one dot may represent 50, or perhaps 100, sheep. The dots should all be the same size.

downland An area of hilly pasture, especially in Australia and New Zealand.

downs Open, rolling chalk uplands, especially in southern England. Mainly treeless grassland with thin soil, downs are used traditionally as permanent pasture for sheep, although some areas have been plowed up to grow cereals, especially barley.

drainage The removal of water, whether by natural means, such as rivers, or by artificial means, including drainage pipes and channels. Natural drainage includes the **throughflow** and **infiltration,** as well as the channeled flow in rivers.

drainage basin SEE **catchment area** [def. 1]

drift 1 Material which has been transported and then deposited by ice. It may consist of clay, sand or larger gravels, or a combination of all of these. It can also include **fluvio-glacial** materials which have been transported by **meltwater. 2** A broad, slow-moving ocean current, for example, the North Atlantic Drift. **3** SEE **continental drift**

drift mine A type of mining in which a sloping tunnel is dug into the ground to give access to the mineral that is to be extracted.

drizzle A light form of rain consisting of small droplets which are only just heavy enough to fall. It is generally associated with **stratus** cloud.

drought A prolonged period of dry weather. There are different definitions in different parts of the world. The degree of aridity of an area will depend on the temperature and the amount of evaporation, as well as on the amount of rainfall. Drought is experienced annually in some regions, such as parts of Ethiopia and the

DRAINAGE

dendritic

parallel

pinnate

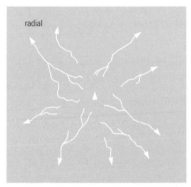

radial

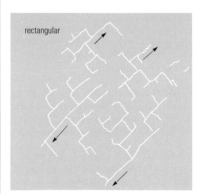

rectangular

trellis

Dendritic and pinnate drainage carry water toward a single point from which it will continue its journey to the sea. In parallel and trellis dainage, water travels in one direction, but not toward a single point. In radial and rectangular drainage, water moves away from an area in many directions.

Sahel in Africa. Some regions of India may have sufficient rainfall for crop growing some years, but then may have a year with less rain than is needed to grow crops. This occasional drought can be more serious than the droughts which occur every year in some regions. The severity of a drought will vary from country to country. A drought in some countries may mean that restrictions will be placed on watering lawns or using hoses for washing automobiles, but a drought in other countries may mean that there is starvation. SEE ALSO **absolute drought**

drowned coastline A coastal strip which has been submerged under the sea, either because the sea level has risen or because the level of the land has sunk. Valleys become flooded, creating **rias**, and hills may become islands.

drowned valley A submerged valley, which has been flooded by the sea or a lake. Sometimes valleys are flooded when dams are built. Lakes created in this way may not only drown farmland but also previously populated areas.

drumlin An elongated mound of glacial **till**. About 1,500 ft [500 m] in length and 30 ft to 60 ft [10 m to 20 m] in height, a drumlin consists of unstratified glacial debris. Drumlins usually occur in groups, called "swarms," which are sometimes described as "basket-of-eggs topography" because of their appearance. They were probably deposited by a glacier or ice sheet which had stopped advancing and was beginning to melt, the movements of the ice producing the characteristically elongated shape.

dry adiabatic lapse rate The rate at which temperature is lost when dry air is ascending, normally about 1°F per 200 ft [1°C per 100 m].

dry farming A type of farming without irrigation in an area of low rainfall. It is most commonly associated with wheat growing, and examples are found in parts of the USA, Australia and southern Africa.

In some dry-farming areas the seeds are sown more thinly than usual. This enables each seed to draw water from a wider area. The yield per acre is low, and so dry farming is extensive and farms are normally very large. Crops may be grown in alternate years only, so that the land can be rested. During the **fallow** year some of the rainfall will be retained in the soil, boosting the following year's total and enabling a satisfactory crop to be grown.

dry site A small elevation in a predominantly wet area. In some marshy environments, for example, there are small hills that are suitable for settlements.

dry valley A river-formed valley which no longer contains a stream. In limestone areas there are many dry valleys. They often occur where streams which used to flow on the surface have opened up a **joint** by **solution** weathering and have then disappeared below the surface to flow in an underground channel. In other areas, some are dry valleys are due to lowering of the **water table**. Others are the result of **periglacial** activity. During the last glacial phase, if the climate was of the **tundra** type, rocks were frozen solid by **permafrost** and rivers had to flow on the surface and eroded valleys. When the climate warmed up in postglacial times and the permafrost had gone, the water sank into the **porous** rock, leaving streamless valleys on the surface. In some dry valleys, a stream may appear after a period of rain.

dune SEE **sand dune**

dust bowl An area which has been changed from a grassy landscape into near desert. If grassland is plowed and the land left bare, wind can blow away the **topsoil** and rain will also wash away soil and possibly form gullies. This eroded landscape is bare and useless for farming. It is also possible to expose the soil by overgrazing; if there are so many animals that they remove all the vegetation, wind and rain will erode the soil. There are dust bowls in many places, including eastern and southern Africa, and the **Sahel** in central Africa. The term was originally applied to the dust bowl of the Great Plains in the USA. It is very difficult to repair the damage once the top-soil has been removed, and so in the USA great efforts have been made to prevent the creation of further dust bowls. Some of the drier grasslands are left unplowed, cover crops are planted to protect the soil, and there has been some **afforestation**.

dust devil A small, localized **whirlwind**, which moves as a pillar of dust across a landscape. Dust devils are particularly common in dry, sparsely vegetated areas, where strong currents are created by the extreme heating of the land surface.

dyke SEE **dike**

dynamic equilibrium A state in which balance is maintained in spite of continual change. For example, a slope where the rate of **weathering** of the rock is balanced by the rate of removal of the weathered material.

E

Earth One of the nine planets in the Solar System and the only one with known human life. It has an oblate spheroid shape (i.e. a flattened sphere) and revolves around the Sun once a year in an "counterclockwise" direction, tilted at a constant angle of 66.5°. Its land surface covers 29.2% (58 million sq miles [149 million sq km]) and the water surface covers the remaining 70.8% (140 million sq miles [361 million sq km]). Above the surface is the **atmosphere,** which protects and provides energy for the Earth's inhabitants. The Earth itself is made up of several layers: the crust, the mantle and the core, with a temperature at the center of the Earth of about 9,000°F [5,000°C].

earth movement A movement of the Earth's crust due to activity beneath it. This includes sudden movements (earthquakes, volcanic eruptions) and slow movements (uplift, folding).

earthquake A tremor below the surface of the Earth which causes shaking to occur in the crust. The shaking will only last for a few seconds, but widespread devastation can result. The earthquake movements are caused by the movements of crustal plates, and when the shock takes place, three different waves are created; the waves are called P, S and L (primary/push, secondary/shake and longitudinal/surface). The P and S waves come from the "seismic focus," which is the point at which the earthquake originated, and travel up to the surface, where they cause shaking. The P and S waves travel along the surface of the Earth as L waves. The point on the surface

▶ **Earth** The 4,600 million years since the formation of the Earth are divided into four great eras, further split into periods and, in the case of the most recent era, epochs. The geological map of the Earth shows the relative dates of the Earth's continents and landscape features.

- pre-Cambrian shields
- sedimentary cover on pre-Cambrian shields
- Paleozoic folding
- sedimentary cover on Paleozoic folding
- Mesozoic folding
- sedimentary cover on Mesozoic folding
- Cenozoic folding
- sedimentary cover on Cenozoic folding
- intensive Mesozoic and Cenozoic volcanism
— principal faults
— oceanic marginal troughs
— midoceanic ridges
— overthrust fault

directly above the seismic focus is called the "epicenter," and the greatest amount of damage generally occurs near here. A large earthquake is generally followed by smaller shocks, known as "aftershocks." If the earthquake occurs beneath the sea, waves can be formed in a way similar to the ripple created on a pond by throwing a stone into the water. These waves are called "tidal waves," but are not connected with tides; a better name for them is the Japanese word **tsunami**. Tsunami can travel across oceans at up to 250 mph [400 km/h], causing damage thousands of miles away from the earthquake's source. Earthquakes in Chile have caused tsunami which have hit the coastline of New Zealand. One of the largest recorded tsunami was over 30 ft [10 m] high when it hit Java and Sumatra in Indonesia, and the water penetrated hundreds of feet inland. This occurred after the massive eruption of Krakatoa in 1883. The main earthquake regions are found along the edges of the Earth's plates, especially

MAJOR EARTHQUAKES SINCE 1975			
Year	Location	Magnitude	Deaths
1976	Guatemala	7.5	22,778
1976	Tangshan, China	8.2	255,000
1978	Tabas, Iran	7.7	25,000
1980	El Asnam, Algeria	7.3	20,000
1980	S. Italy	7.2	4,800
1985	Mexico City, Mexico	8.1	4,200
1988	N.W. Armenia	6.8	55,000
1990	N. Iran	7.7	36,000
1993	Maharashtra, India	6.4	30,000
1994	Los Angeles, USA	6.6	57
1995	Kobe, Japan	7.2	5,000
1995	Sakhalin Is, Russia	7.5	2,000
1996	Yunnan, China	7.0	240
1997	N. E. Iran	7.1	2,500
1998	N. Afghanistan	6.1	4,200
1998	N. E. Afghanistan	7.0	5,000
1999	Izmit, Turkey	7.4	15,000
1999	Taipei, Taiwan	7.6	1,700
2001	Gujarat	7.7	18,600

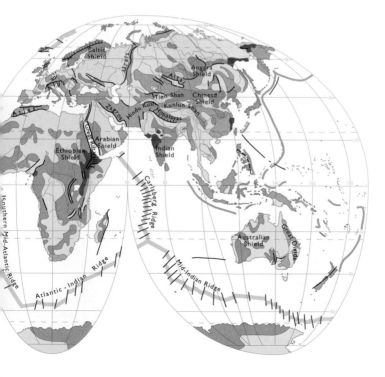

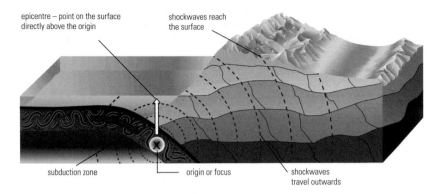

epicentre – point on the surface directly above the origin

shockwaves reach the surface

subduction zone

origin or focus

shockwaves travel outwards

▲ **earthquake** Earthquakes occur when crustal plates move against each other. Here, one plate moves under another a "subduction zone". The movement of the plates causes the earthquake shock, spreading outward from the origin. The epicenter is the point on the surface directly above the origin. In this case, the epicenter is in the sea, so a tsunami may occur. Damage on land occurs when the shockwaves reach the surface, with the amount of damage lessening as distance from the epicenter increases.

on the perimeters of the Pacific. Earthquakes are usually measured on the **Richter scale**. SEE ALSO **plate tectonics**

easting The north–south grid lines on some maps, such as military ones. They are given first when quoting a map reference.

ebb tide The retreating or falling tide.

eclipse The blocking out of the Sun's light by the intervention of a celestial body between the source of the light and the recipient. When the Moon is in a direct line between the Earth and the Sun, it blocks some of the sunlight which would normally reach the Earth; this is an eclipse of the Sun (or "solar eclipse"). If the Earth is in a direct line between the Sun and the Moon, the Sun's rays will not reach the Moon; this is an eclipse of the Moon (or "lunar eclipse").

ecology The science which studies living organisms in relation to their **environ-**

ment and to other living things. It is concerned with the interrelations between the different components of the **ecosystem**.

economics A broad term summarized as the study relating to the production and distribution of material wealth and to human attitudes to working within the limits imposed by the availability of resources.

economies of scale The reductions in unit cost which result from large-scale, as opposed to small-scale, production. By means of mass production and automation, it becomes possible to produce goods more cheaply. The aim is always to reduce the unit costs of the items produced. The effects are evident in large, integrated automobile manufacture.

ecosystem A community of living organisms (plants and animals) and the location or **environment** in which they live. An ecosystem can be very large (for example, the Earth) or very small (for example, an oak tree or a lawn). All the elements of the ecosystem, whatever its size, will be inter-related. There will be flows of energy and nutrients, and a change in one part of an ecosystem may affect several others. Left to nature, an ecosystem will achieve a balanced state, in which plants and animals live together. A sudden change or a disaster, such as very heavy rain or a prolonged drought, will affect some of the plants and animals, but gradually they will recover. If an oak tree was blown down, the elements of the oak system would contribute to another system. In recent years, many ecosystems have been seriously affected by human activity. There are numerous occasions when human influence has been harmful, or even disastrous; for example, cutting down trees, plowing up land, and causing soil erosion. There are major problems which can affect large areas; very large ecosystems are damaged by **acid rain**, and it is possible that the world ecosystem may suffer as a result of the **greenhouse effect**.

edaphic Relating to soil conditions. The edaphic factors are those which influence the growth of plants, such as soil texture, mineral content and soil moisture.

eddy A roughly circular movement within a current of air or water, in an opposite direction to that of the main flow. For example, if water in a river comes across an obstruction.

EC SEE **European Union**

edge SEE **grit**

▼ **eclipse** A solar eclipse occurs when the Moon passes between the Earth and the Sun. If the Moon is in a direct line between the observer and the Sun, the observer sees a total eclipse. Otherwise, the observer sees a partial eclipse. While an observer at one place on Earth sees a total eclipse, an observer elsewhere may see a partial eclipse. A lunar eclipse occurs when the Earth casts a shadow on the Moon.

effluent Flowing out. This may be the outflow of sewage or industrial waste, or a stream that flows out of another stream or lake.

elbow of capture A right-angled bend in a river. It is shaped like an elbow because river capture has taken place. Upstream of the bend is the captured stream and downstream of the bend is the capturing stream. SEE ALSO **river capture**

El Niño A warm ocean current found off the coast of South America in the eastern Pacific Ocean, which occurs every six to 14 years and replaces the normally cold currents found there. It is thought to coincide with a failure of the southeast **trade winds** and therefore a failure of the cold currents driven by them. By warming the nutrient-rich waters, many of the fish and plant species are killed.

eluviation The removal of material from the **A horizon** of a soil. Chemicals in solution or small particles of solid matter can be taken down to the **B horizon** by percolating water. The transported material will accumulate in the illuvial layer.

employment structure The classification of different types of work. Workers are employed in the **primary**, **secondary**, **tertiary** or **quaternary** sectors. Tertiary and quaternary are often classified together under the heading of tertiary. Primary employment is in agriculture, forestry, mining and fishing. Secondary employ-

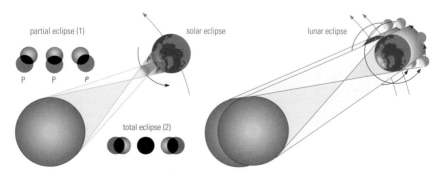

partial eclipse (1)   solar eclipse   lunar eclipse

P   P   P

total eclipse (2)

EL NIÑO

normal year – Walker Circulation Cell

El Niño event

In a normal year, surface water flows eastward from South America under the influence of trade winds. Near the coast, cold water rises and spreads westward. In the western Pacific, the sea surface is warm and warm air rises, causing rain. Some of this air spreads eastward and descends over South America, causing winds that blow westward. In an El Niño year, the currents reverse. The eastward-moving warm currents extend much further eastward and the trade winds weaken. The rising of cold water off South America is reduced and surface water temperatures rise. Warm air rises here, causing rain, and descends over Southeast Asia.

ment is in the manufacturing industry. Tertiary employment is in retailing, services and administration, and quaternary employment is in the provision of information, professional advice and expertise. In less developed communities, a high percentage of the workforce will be in agriculture, but as countries become more advanced and wealthier, employment in agriculture declines, the manufacturing industry grows, and eventually tertiary and quaternary activities begin to dominate employment. By examining the employment structure figures, it is possible to determine whether a country belongs to the developed or less developed nations.

enclave A piece of territory lying completely within foreign territory. For example, West Berlin was an enclave within East Germany between 1945 and 1990.

enterprise zone British term for an area where special grants and tax advantages are available to anyone creating employment. The government may offers

incentives to encourage firms and companies to set up in areas of high unemployment. Most enterprise zones are in inner city areas.

entrenched meander SEE **incised meander**

entrepreneur A person who wholly or partly undertakes and develops an enterprise, particularly a commercial one, usually at some financial risk to himself.

environment One's surroundings, conditions and circumstances considered as a whole. It may be a natural physical environment or an artificial urban environment. Much of the countryside is also a man-made environment, where agriculture has changed the appearance of the landscape. The soil, vegetation and buildings are all part of the environment; and pollution, survival of animal species, soil erosion, deforestation and many other topics are relevant to the continuation of an environment. Conservation groups attempt to influence the ways in which the environment is being changed by humans, and forest parks, nature reserves, etc., are being created in many places in order to protect parts of the environment.

environmental hazard Sources of danger which arise in the environment. These may be natural, such as floods, droughts or earthquakes, or man-made, such as pollution, oil spills or acid rain.

environmental lapse rate SEE **lapse rate**

epicenter SEE **earthquake**

epilimnion The upper, warmer layer of water in a lake or the sea, into which light may penetrate for **photosynthesis** to occur. This layer may be also disturbed by wind and currents.

Equator An imaginary line around the Earth which represents the 0° line of latitude and encircles the broadest part of the Earth. It is 24,901 miles [40,076 km] long and is the only line of latitude which is a **great circle**, although all lines of longitude are great circles.

equatorial rain forest SEE **tropical rain forest**

equilibrium A state of balance. SEE ALSO **dynamic equilibrium**

equinox The time of year when day and night are of equal length at all points on the Earth. The "spring equinox" (or "vernal equinox") is on 21 March, and the "autumnal equinox" is on 22 September. On both of these dates the Sun is directly overhead at noon on the Equator, and both the North Pole and South Pole will be receiving light.

erosion The process by which the Earth's surface is worn away, principally by water (rivers, ice and the sea) and wind. These agents, together with **weathering**, are responsible for causing natural changes in the

landscape; they are also responsible for **deposition**, which can create new features.

erratic block A rock which has been transported from its place of origin. Ice-sheets and glaciers carry large quantities of rocks, sometimes for many miles. Most deposition takes place as the ice is melting, and frequently the rocks are deposited on a completely different type of rock. Erratics are found in Canada, the USA, Britain, Switzerland and many other countries. In parts of New England and Minnesota in the USA they are so numerous that they hinder the use of machinery on the farms.

escarpment A steep slope (or "scarp") on one side of a hill, on which the other side is a gentle slope (or "dip slope"). The escarpment may be the result of faulting, or, more frequently, caused by tilting of the strata, followed by differential erosion. SEE ALSO **cuesta**

esker A long narrow ridge formed by a subglacial stream. There are often streams flowing beneath glaciers and ice sheets, and they carry deposits of sand and gravel. The rivers flow in tunnels and may deposit large quantities of sand and gravel on their beds. The deposits build up and are left behind when the ice melts and the rivers move to other channels. Eskers are usually 30 ft to 80 ft [10 m to 25 m] high, and 15 ft to 80 ft [5 m to 25 m] across; some extend for many miles, especially in Finland and northern Canada. In Finland there are eskers which form boundaries to lakes, and some are used as routes for roads.

estuary The mouth of a river where tidal effects can be seen. In nontidal seas, only the effects of salt water will be evident. Most estuaries are funnel-shaped, becoming wider nearer the sea. They may be the result of subsidence and flooding. At low tide, large expanses of mud or sand are visible in many estuaries, and **deposition** may be occurring. Some estuaries contain ports, and they provide good sheltered positions. Dredging may be necessary to keep the channel open, but if the estuary narrows,

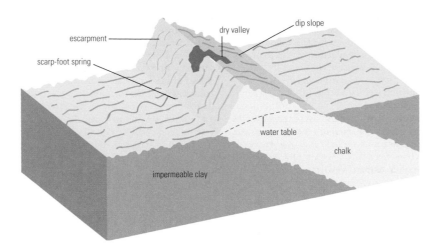

escarpment
dip slope
dry valley
scarp-foot spring
water table
chalk
impermeable clay

▲ **escarpment** The escarpment shown here has resulted from the tilting of a chalk stratum. The spring at the foot of the escarpment arises because the water table in the permeable chalk is higher than the clay surface but cannot penetrate the impermeable clay. The dry valley was created by water, but is now well above the water table.

there may be a scouring effect caused by the river flow.

ethnic cleansing A campaign inflicted on one ethnic group by another to force them from their homes or areas, usually by violent methods and often including imprisonment, torture and death. The term has been used to describe the Serbian campaign to force Muslims from their homes during the war in the former Yugoslavia that began in 1992.

ethnic group A group living within a larger society which has its own distinctive customs and culture. Such a group may be the result of former immigration and may settle in large towns, in areas known as **ghettos**.

European Community (EC) SEE **European Union**

European Union (EU) The European Union evolved from the European Community (EC) in 1993. The original body, the European Coal and Steel Community (ECSC), was created in 1951 following the signing of the Treaty of Paris. The 15 members of the EU – Austria, Belgium, Denmark, Finland, France, Germany, Greece, Ireland, Italy, Luxembourg, Netherlands, Portugal, Spain, Sweden and the UK – aim to integrate economies, coordinate social developments and bring about political union. These members, of what is now the world's biggest market, share agricultural and industrial policies and tariffs on trade. Plans to expand EU membership include countries in Eastern Europe and on the Baltic Sea.

eutrophication The process whereby a body of fresh water becomes more fertile ("eutrophic"), often as a result of pollution from sewage, industrial waste or agricultural fertilizers. An **algal bloom** may follow this increased fertility.

evaporation The process by which a liquid (or solid) changes to a vapor or gas, at a temperature below the boiling point of the liquid. Evaporation will take place at the surface of the liquid.

evapotranspiration The total loss of moisture by evaporation from water surfaces and the soil, and by transpiration from plants. It is quite difficult to measure evapotranspiration because of the interrelationships between precipitation, runoff, evapotranspiration, and the changes in soil moisture. Precipitation = runoff + evapotranspiration ± the changes in soil moisture storage.

evergreen trees Trees which retain their foliage throughout the year, most frequently conifers, except in equatorial regions where, because of the lack of seasonal variation in climatic conditions, many deciduous trees become evergreen. As leaves are shed, so new foliage is produced in a continuous canopy. The trees are able to grow new leaves each month because the climate is hot and wet. SEE ALSO **coniferous forest**, **deciduous forest**, **taiga**

exfoliation A weathering process in which small pieces flake off a rock, rather like skins off an onion. It is most active in deserts and is the result of extreme changes in temperature between day and night. The rocks heat up during the day and expand. As they cool at night they contract. Repeated expansion and contraction weakens the rock, especially if there is water in the rock, as this will increase the amount of expansion. In deserts the presence of **dew** is a contributory factor. Rocks are rounded off by exfoliation; there may be small piles of rock debris beside the larger rocks. **Freeze-thaw** activity is similar to exfoliation, but there is a larger amount of expansion involved because water increases by 9% in volume when it freezes. ALSO CALLED **onion weathering**

exosphere The atmosphere's upper layer, with a lower boundary at an altitude of around 370 miles [600 km] and no clearly defined outer boundary. It is mainly composed of helium and hydrogen in changing proportions. The amount of helium decreases with increasing altitude, and above 1,500 miles [2,400 km] the exosphere is almost entirely composed of hydrogen. SEE ALSO **atmosphere**

export A good or service sold to another country with payment by money or other goods and services (**imports**). There are two sorts of export: invisible, which consist of payments for services, such as banking, insurance, shipping and tourism; and visible, which consist of actual goods, such as raw materials and manufactured goods. SEE ALSO **invisible exports**

exotic stream The name given to a few large rivers, such as the Nile in Egypt or the Indus in India, which can flow across major desert regions because they are fed by rain falling on mountains hundreds of miles away. There are more than 50 such streams flowing across the Atacama Desert in Peru. SEE ALSO **intermittent stream**, **perennial stream**

exponential (*of population*) Showing a rapid increase. The graph of world population has shown a dramatic growth during the last 40 years. This increase has been due to improvements in medicine which have reduced the **death rate** but have not been accompanied by a corresponding decline in the **birth rate**. This rapid increase is described as "exponential growth."

exposed coal field A coal field in which coal seams reach or almost reach the surface. In exposed coal fields it is possible to mine by the open-cast method, which is cheaper than shaft mining. However, most exposed coal fields were utilized in the early days of mining, since coal was most accessible. SEE ALSO **concealed coal field**, **open-cast mining**

extensive farming Large-scale agriculture in which large areas of land may be used, but in which the return per acre is relatively small. Sheep farming on hills or in semiarid areas is a good example of this. Most extensive farming relates to pastoralism, but there are some areas of extensive arable; for example, the dry-farming areas in parts of Australia.

extrusive rocks Igneous rocks which were forced out in a molten state (as magma) at the Earth's surface, where they cooled quickly and then solidified as

lava. Because they cooled rapidly, extrusive rocks contain very small crystals, which may even be too small to be seen by the naked eye. If, however, magma cools and solidifies beneath the surface, cooling will be much slower and larger crystals will form. The rocks formed in this way are called "intrusive rocks." SEE ALSO **magma**

eye The calm, still area at the center of a hurricane.

F

factors of production The requirements for production: usually seen as capital, labor and land. Capital is "fixed," as in the actual buildings and machinery, and "variable," such as raw materials and components. Labor may be unskilled, semi-skilled or skilled; and land may be an actual raw material, as in mining, or a location for buildings.

factory farming Intensive livestock farming, in which cattle or pigs are kept permanently indoors and fed a carefully regulated amount of food each day. The animals are not allowed to walk about very much, as too much exercise would use up energy and slow down their growth rate. It is a very commercial type of farming, which aims to produce a good-quality product as quickly as possible. When chickens are reared in this way it is called "battery farming."

Fahrenheit A temperature scale which is used in English-speaking countries. It remains the standard scale in the USA, but has been largely replaced elsewhere by **Celsius**. Freezing and boiling points of water on the Fahrenheit scale are 32° and 212° respectively.

fallow A term denoting farmland which has been left unseeded for a season or more in order to rest it. The land may be tilled and the weeds may be killed off during the fallow period. Some fields may be left fallow for a season before planting a late or winter crop. In some tropical areas, "bush fallow" is practiced; with this system the land may be left fallow for several years.

family planning Birth control by the use of contraceptives. This is already common practice in many advanced countries, and is now an urgent requirement for most **Third World** countries, where many of the social and economic problems result from overpopulation in relation to the available resources. SEE ALSO **birth control**

famine An extreme and general shortage of food. Famines have been an occasional problem for hundreds of years, generally occurring as a result of variations in the amount of rainfall, or natural disasters such as volcanoes and hurricanes. Too much rain can cause flooding and delay planting of crops, or destroy crops which were already growing. Too little rain will prevent crops from growing, and so the harvest will be small. Although there is now more food grown in the world and more help available for people who are starving, populations have grown so rapidly that famines occur more frequently than in the past. In some countries, such as Ethiopia, famine is a common occurrence.

farming system Any type of farming with its own distinctive characteristics, and inputs and outputs. In England, a dairy farm in Cornwall is an example of a farming system, while a cereal farm in East Anglia is a different farming system, with more machines, bigger fields and different products. A shifting cultivator in the Amazon basin in South America is a totally different system, as is a Lapp reindeer herder in northern Scandinavia; there are many other examples all over the world.

farmscape A landscape which has been created by the farmers. On the borders of England and Wales there are many small fields with numerous hawthorn hedges, which create a very different landscape to the US and Canadian prairies. Farming methods determine the appearance of the landscape.

fault A crack or fracture in the rocks of the Earth's crust with an associated movement of the strata on either side. The movement will be slow and quite small, only a few inches, though fault movements often continue for thousands of years. In such cases, uplift or downthrow of hundreds of feet is possible. Faulting is caused by **plate tectonics**, when movements in the Earth's crust create stress and tension in the rocks, causing them to stretch and crack. Vertical movements cause "normal" and "reverse faults," and horizontal movements cause "tear faults." The vertical change of height on opposite sides of a fault is called the "throw," and the horizontal movement is called the "heave." Faults often occur in groups and if two or more roughly parallel faults cause a block of land to rise, it is called a "horst," or **block mountain**. If the land sinks between two or more parallel faults, this would create a "graben," or **rift valley**.

fault plane The surface along which faulting has occurred. In the case of a normal fault, the fault plane will be vertical, or it will be inclined so that the downthrown side is on the dip side of the fault plane.

fault scarp A scarp which is located along the line of a fault. The north face of the San Jacino Mountains in California, USA, is an outstanding example.

favela (*in Portugal and Brazil*) A spontaneous or shanty settlement which has grown up in a large city, such as São Paulo in Brazil. SEE ALSO **shanty town**

feed Food for animals. It may be hay, root crops, alfalfa, oil seed cake, etc.

feldspar Any of a large group of minerals which are silicates of aluminum, with some potassium, sodium, calcium or barium. Almost half the Earth's crust is made of feldspars. Common varieties include "orthoclase," which is often whitish in color, and "plagioclase," which is often pinkish; both varieties can be found in **granites**. ALSO CALLED **felspar**

fell An upland in northern England, especially in the Pennines and Lake District. The fells are areas of rough grazing often used for sheep in the summer. Some of the fells are common grazing and some are important grouse moors.

felspar SEE **feldspar**

FAULT

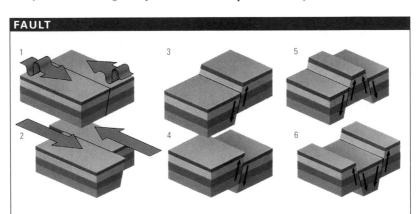

The Earth's crust is subjected to enormous forces, and the stress creates faults. In a tear fault (1), the stresses cause horizontal movement. The forces build up until they are released in a sudden movement (2), often causing earthquakes. In a normal fault (3), the rocks are pulled apart, causing one side to slip down along the plane of the fault. In a reverse fault (4), the rocks on either side of the fault are forced together. One side rises above the other along the fault plane. In a horst fault (5), the central section is left protruding due to compression from both sides or the sinking of the bracketing rock. A rift valley (6), has a sunken central section, formed either by compression or by the outward movement of the two valley sides.

fen A flat low-lying area which is marshy. Regions of fen occur in many river valleys where there used to be lakes, and near the coast where land has been reclaimed from the sea by silting. Some fen soils are very peaty and acidic, but others are alkaline in character.

Ferrel's law The law that, because of the rotation of the Earth, all moving bodies on the Earth's surface will be deflected to their right in the northern hemisphere and to their left in the southern hemisphere. This is especially apparent with reference to winds, but it can also affect ocean currents, and people; for example, people who are lost in fog or in a desert walk around in circles as a result of Ferrel's law.

fertilizer Any material used to improve the fertility of the soil. Natural fertilizers are obtained from animal manure, rotting vegetation, fish bones, etc. Many artificial fertilizers are now produced by chemical industries, using potash, phosphates and nitrogen.

fertility The condition of being fertile, i.e. the ability to produce live-born children. The "general fertility rate" gives the number of births in a year per 1,000 women of reproductive age (usually 15–45). Fertility rates may vary in different areas due to various factors, such as access to birth control, income, death rate, etc.

fetch The extent of sea across which winds can blow. If there is a large expanse of open sea, it is likely that high **waves** can be expected. For example, waves which come in from the Atlantic Ocean are likely to be larger than waves which form in the narrow parts of the English Channel.

finance capital SEE **capital**

finger lake A long narrow lake situated in a glacial **U-shaped valley**. As the glacier moved downhill, deepening and widening the valley, the irregular movements of the ice along the valley floor would cause uneven erosion. When the ice melted, hollows would be left behind in rock basins. The steps at the end of each basin would form a dam, and occasionally there would be deposits of glacial drift to add to the damming effect. ALSO CALLED **ribbon lake**

fiord SEE **fjord**

firn (*German*) Granular snow, halfway between snow and ice, which is subsequently compressed and compacted to form ice. ALSO CALLED **névé**

First World Those countries, in the developed world, with some form of a capitalist, free-market economy; including North America, Western Europe, Australasia and Japan. Recent political changes (since 1990) in Eastern Europe, once part of the **Second World**, has meant the division between the First and Second Worlds has become less distinct. SEE ALSO **Third World**

fissure A cleft or crack in a rock, especially one which is the site of a **fissure eruption**.

fissure eruption A volcanic eruption in which lava pours out of a fissure and continues to flow for a few hours, sometimes days. **Basic lava** is fluid and free-flowing, and can travel across large areas of countryside. There may be several eruptions along the same fissure line. When this happens, large deposits of lava can accumulate, and there are examples of plateaus which have been formed in this way, such as the Snake/ Columbia plateau in the northwest USA, parts of the Deccan plateau in India, and the Antrim Mountains in northeast Ireland.

fixed capital SEE **capital**

fjord An inlet of the sea which was formed by glacial erosion. It is really a flooded **U-shaped valley**. Fjords are generally narrow and steep-sided, and the water can be hundreds of feet deep. The bed of the fjord is often undulating because of the irregular curve of glacial erosion. Sometimes there is quite shallow water at the seaward end, because less erosion took place near the end of the glacier. Large examples of fjords can be seen in Norway, southern Chile and South Island, New Zealand, and there are some examples in northwest Scotland and northwest Ireland. ALSO CALLED **fiord**

flash flood A rapid flood, often quite dramatic, caused by heavy rain followed by a rapid runoff. Flash flooding is most likely to occur in deserts, where there is no vegetation to slow down the rate of runoff, but it can also occur in wetter areas. Heavy rain, probably the result of a thunderstorm, may fall several miles away from the site of the flood. Routes followed by flash floods are

sometimes well known, and in the USA and Israel, for example, there are places where signposts give warnings of the likelihood of flash floods.

flint A hard rock which consists mainly of the mineral silica. Flint was formed from the remains of dead sea creatures whose bodies were made of silica. It is often found in chalk, which is also derived from dead sea creatures but of a different kind. Layers of flint commonly occur in the chalklands of southern England. They are a nuisance in fields, because they are so hard and interfere with machinery. They are often used

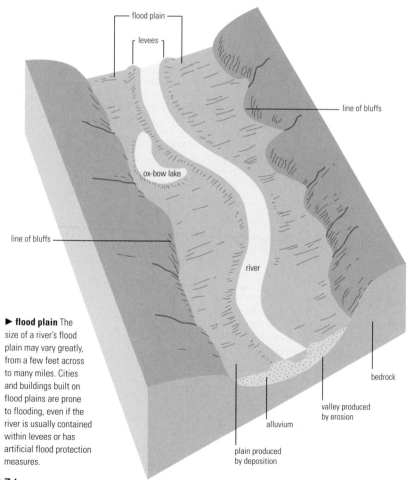

► **flood plain** The size of a river's flood plain may vary greatly, from a few feet across to many miles. Cities and buildings built on flood plains are prone to flooding, even if the river is usually contained within levees or has artificial flood protection measures.

for building purposes, as they are very resistant to **weathering**. Many churches in chalkland regions have been faced with pieces of flint.

floating capital SEE **capital**

flooding Flowing of water over land which is not usually submerged. Very heavy rainfall, prolonged steady rain or melting snow can be the cause of flooding. Some river basins, for example in granite areas, have a rapid runoff and water quickly reaches the river; in limestone and chalk areas, the water passes slowly through the rocks and the danger of flooding is reduced. In **porous** rocks, the peak flow in rivers often comes a day or more after even the heaviest rainfall, whereas in **impermeable** areas the peak flow is only hours after rainfall. A good cover of vegetation, especially with woodland, delays the **throughflow** of water. If the vegetation cover is removed, there may be a much greater danger of flooding. Many parts of the world, including Nepal, Thailand and the Amazon basin in Brazil, now suffer from more serious flooding than in the past because water runs off far more quickly as a result of extensive tree felling in these areas.

flood plain A lowland alongside a river which has been built up by the deposition of **alluvium** in times of flood. Many flood plains are only a few feet in extent, but on some rivers, such as the Mississippi in the USA and the Amazon in Brazil, the flood plain can stretch for several miles. The flood plain of the River Huang He (Hwang Ho) in northern China is over 100 miles [160 km] wide, and there are large **levees** alongside the river. When the river floods, the water cannot always return to the main channel and so a new channel may be created; in this way, the Huang He has frequently changed its course.

flow line A type of diagram on which a line represents the movement of goods or people. The width of the line is proportional to the amount of movement which is taking place. To show the international trade in wheat, for example, arrows could be marked on a world map; there would be thick arrows from the USA and Canada, and also an arrow from Australia to show the main exporters. The arrows would then split up to indicate the main wheat importers.

fluvial Relating to a river or a stream.

fluvioglacial Relating to a river which has been formed by **meltwater** from a glacier or ice sheet. There are often rivers which flow beneath glaciers; they may form **eskers** and **kames**, which are fluvioglacial deposits. At the end of an ice sheet, where melting is taking place, there is likely to be a large quantity of water flowing across the plain or down the valley. At first it may be a sheet flow, but gradually it will become a channeled flow. It will pick up and transport particles of sand and perhaps gravel which have been dumped by the ice. Coarse particles will be dropped first, but the fine clays may be carried considerable distances. The deposition of the sediments will be according to weight and size, and so sorting inevitably takes place; fluvioglacial deposits are always sorted, and often stratified, too. The bulk of fluvioglacial deposits are found in the **outwash plain**, which is quite close to the downstream end of the ice.

fog A mass of small water droplets floating in the air which make visibility less than 0.5 miles [1 km]. If the visibility is more than 0.5 miles [1 km], it is **mist**. Fog is caused by water vapor condensing as a result of the air becoming cooler. It is most likely to form in settled anticyclonic conditions. After a hot and sunny day, the air is warm. When night falls the air cools, and so is less able to hold water vapor. The water vapor condenses to form millions of tiny droplets. The coldest air is likely to roll downhill into valleys, which is where fog frequently forms. Fog is also quite likely to form first near rivers or lakes, where there is a greater supply of moisture. If a wind develops, the turbulence will clear the fog. Fog will also clear in the heat of the sun during summer days. In winter, however,

FOLDING

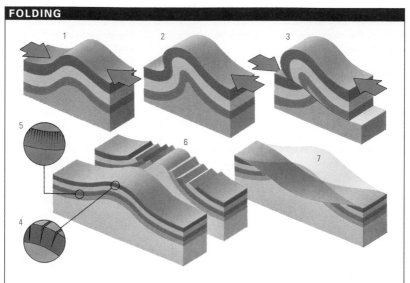

Tectonic forces warp the Earth's surface and are powerful enough to bend strata of rock. An anticline (1) is created when the rock is pushed upward. When the rock is forced down, a syncline is formed. If the force continues an overfold (2) forms. If the strata are warped too much, they can in effect snap (3) making an overthrust fold or nappe, which is both a fold and a fault. At the top of an anticline, the rocks are "stretched" (4), while at the base of a syncline they are compressed (5). The extra joints and small cracks at the top of the anticline makes it more prone to weathering and erosion (6). If this process continues over many thousands of years, an anticline can be eroded to form what appears to be a syncline (7).

it can persist throughout the day. "Hill fog" is really low cloud, and is associated with low pressure. It may persist for several hours, but will not affect low-lying areas.

föhn (*German*) In Europe, a warm dry wind which blows down the valleys on the north side of the Alps. It is associated with the movement of a **depression** to the north of the Alps. The low pressure of the depression draws in air from the south, which rises over the Alps; as it ascends, condensation occurs and there is rainfall. The ascending air loses heat at a slow rate because the air is wet. Once over the range of mountains, the air is dry and so gains heat at a more rapid rate. By the time it descends the valleys on the other side, it has become a very warm dry wind. The *föhn* occurs most frequently in spring, when it is useful for melting the snow. If it occurs in the fall it can help with ripening the grapes, as an increase in temperature of about 20°F [10°C] is often recorded. A similar wind which blows over the Rocky Mountains in Canada and the USA is called the **chinook**.

folding The process by which rock strata, under pressure as a result of movement of the Earth's plates, buckle and bend. If the compression is fairly gentle and quite even from both sides, the rocks may fold into gentle and symmetrical shapes. If the pressure is uneven, then "asymmetrical" folds will form. In many cases, the folds are pushed right over to form "recumbent" folds and overfolds. Eventually the rock strata may break under the pressure, to form an "overthrust" and finally a "nappe" fold. In Europe, nappes are very numerous in the Alps, where intense folding has taken place; simple folds are found in the Jura

Mountains, and a good example of an asymmetric fold is found in the Brooks Range of Alaska in the USA. SEE ALSO **anticline**, **syncline**

fold mountains Mountains formed as a result of folding, including the Alps, Andes, Himalayas and Rockies. Convection currents beneath the Earth's crust cause the plates to move and compress the sediments; the **sedimentary rocks** are then folded and uplifted to form mountains. In association with the folding there is sometimes volcanic activity, and the central core of some fold mountains contain **igneous** rocks. There have been three major periods of folding and mountain formation during the last 500 million years. The "old" fold mountains were formed in the **Paleozoic** during the Caledonian and Hercynian orogenies; the "young" fold mountains formed in the **Cenozoic**. Caledonian folding occurred at the end of the Silurian period, and formed the mountains of Scotland and Scandinavia. Hercynian (or Armorican) folding came at the end of the **Carboniferous** period and formed the Appalachians of the USA, some of the Welsh mountains, the Harz Mountains in Germany and the Cape Ranges of South Africa. During the Tertiary period at the start of the Oligocene, the Rockies of North America, the Himalayas of Asia, the Alps of Europe and the Andes of South America were formed. Because the processes of erosion have been at work on these mountains for a relatively short time, they are not only the youngest but also the highest mountains in the world. SEE ALSO **geological column**, **geological time scale**, **orogenesis**

food chain A sequence of linked organisms which exist in any natural community, along which energy passes and which represents the nutrition of various species from a simple plant to a prime carnivore (meat-eater). The plants and animals are linked at different levels, joined together in their diet and in their role as a food source. Very basically, the levels rise from a plant to a herbivore to a carnivore.

footloose industries Light industries which have few basic requirements and are therefore able to set up in almost any location that offers an adequate power supply and access to an efficient means of transportation.

forestry The planting, cultivating and management of woodland, usually to provide timber for sale. These areas are often highly protected and well looked after, with the use of fertilizers, special protection from fire and the planting of fast-growing woods such as softwood conifers.

fossil fuel Any of the nonrenewable fuels formed from the remains of dead plants and animals which accumulated in the ground: coal, oil, natural gas and peat. They formed very slowly but are being consumed very quickly, and are likely to be exhausted within the next few hundred years. Burning fossil fuels has led to much pollution of the atmosphere and to the phenomenon known as **acid rain**. Increasing attempts are now being made to clean up power stations, so that they do not give off as many harmful gases.

francophone French-speaking. It is used, for example, by Africans for whom French is a second mother tongue.

free market SEE **capitalism**

free port A port or a zone within a port where no customs duty is paid. Designed to attract more customers and more trade, many free ports, such as Hong Kong, are *entrepôts*, which handle goods to be passed on to other countries. Some free ports are located at airports; for example, Manaus in Brazil. ALSO CALLED **free-trade zone**

freeze-thaw A process in which extreme changes of temperature contribute to the break-up of rocks. During a cold spell, probably at night, the water particles in the rocks freeze. When this happens, the water increases by about 9% in volume, and this places a great strain on the rock. As the temperature rises, the water melts. If this

FOSSIL FUEL

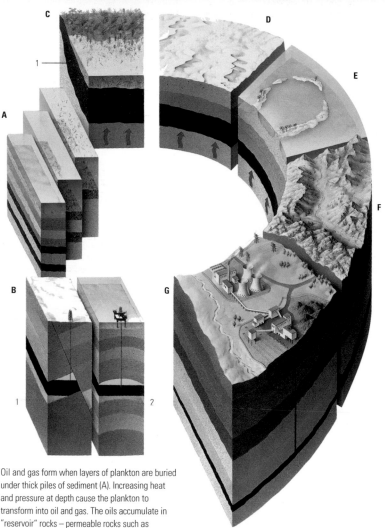

Oil and gas form when layers of plankton are buried under thick piles of sediment (A). Increasing heat and pressure at depth cause the plankton to transform into oil and gas. The oils accumulate in "reservoir" rocks – permeable rocks such as sandstone. To form an oilfield (B), oil or gas must be trapped under impermeable rock in structures such as those created by faults (1) or convex domes (2). Most coal (C–G) began in the Carboniferous period when lush tropical forests grew in extensive swamps (C), and thick layers of peat (1) were laid down. The seas receded and many coastal plains turned to desert (D). Sedimentary rocks, such as sandstones, were laid down. With increasing temperature and pressure, the buried peat began its metamorphosis into coal. Around 150 million years ago (E), the deserts were covered by tropical seas in which limestone was deposited. Around 50 million years ago, mountain formation further buried the coal (F). Metamorphism continued, converting coal into high-grade anthracite suitable for mining (G).

process is repeated day after day for many years, the constant expansion and contraction will gradually break up the rock. Freeze-thaw occurs in mountainous and **tundra** areas. Loose rock fragments resulting from freeze-thaw activity can be seen in the Alps. In glacial areas it is a major contributor to the erosion of back walls of **cirques** and other exposed rocky outcrops.

freezing point The temperature at which a liquid changes to a solid. For example, water changes to ice at 32°F [0°C].

frequency curve A graph which shows the frequency, or number of occurrences, of any chosen item, such as temperatures, answers to a questionnaire, etc. The values are often grouped into classes so that a **histogram** can be drawn. On a frequency curve graph, the horizontal axis gives a scale for the range shown by the variable, and the vertical axis gives the frequency of each variable, either as an actual figure or as a percentage.

frequency distribution The range of values covered by any set of data, generally shown as a **bar graph** or **histogram**. A population pyramid is a frequency distribution diagram. SEE ALSO **age sex pyramid**

friction of distance The restricting effect of distance on human activity and general accessibility. The amount of traffic tends to decrease with increasing distance from a big city such as London, England, or Montreal, Canada. The number of walkers on a sidewalk decreases in proportion to the distance from the nearest parking lot. In trading goods, an increase of distance is likely to increase the cost and therefore decrease the likelihood of selling the commodities. The general formula for friction of distance is:

f   is the volume of movement
d   is the distance
a   is the constant

$$f = a \frac{1}{d^2}$$

There are several useful gravity models which consider the effects of distance. How-

ever, such models cannot really consider the effects of developments in transportation, which speed up the flow in some cases, and cannot take into consideration the peculiarities and idiosyncrasies of people.

Friends of the Earth The largest international network of environmental groups in the world, represented in 58 countries. It is largely funded by supporters, with over 80% of its income coming from individual donations, and the rest from fund-raising events, grants and trading.

front A dividing line between two air masses. SEE ALSO **cold front, occluded front, warm front**

frontal rainfall Rainfall which is formed in a **depression**, where the fronts cause a far larger and more rapid uplift of the air. Large quantities of air rise quite quickly at the fronts and the rainfall can be heavy, though it is unlikely to last for more than a few hours.

frost Frozen droplets of water. When the air temperature falls, generally at night, water vapor may condense. If the air temperature then falls below 32°F [0°C], the water will freeze, forming a covering of "ground frost" on the ground and on any solid exposed objects. If the temperature falls a few degrees below zero, it is possible for the water vapor in the air to freeze to form "air frost," but this is not as common as ground frost; both types of frost are called "hoar frost." Frost is most likely to occur in anticyclonic conditions when the air is fairly calm, and there is little wind or turbulence to mix up the layers of air. The air is cooled by radiation from the ground.

frost heaving The raising of soil and rocks from the ground surface by the freezing and subsequent expansion of soil water.

frost hollow A valley bottom or basin which is prone to frost. In a frost hollow, cold air accumulates, producing lower temperatures than those on adjacent slopes; night minimum temperatures may be as

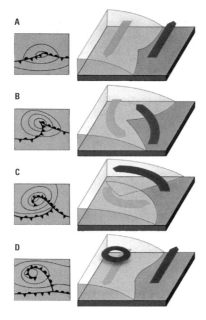

◄ A front forms in temperate latitudes where a cold air mass meets a warm air mass (A). The air masses spiral round a bulge causing cold and warm fronts to develop (B). The warm air rises above the cold front and the cold air slides underneath the warm (C). Eventually, the cold air areas merge, and the warm air is lifted up or occluded (D).

gives rise to **temperature inversion,** with a warmer layer above a zone of colder air. The inversion layer is likely to form **cloud,** which will restrict the upward movement of dust or smoke. **Pollution** is likely to accumulate below the inversion layer, and this is one of the contributory factors to the fogs of Los Angeles in the USA and the former **smogs** of London, England.

Fulani A large group of nomadic and seminomadic pastoralists living in northern Nigeria and parts of adjacent countries. They normally have a permanent village to which they return for a few months each year, but they have to wander with their herds of cattle for several months each year, in order to find pasture.

much as 7°F to 9°F [4°C to 5°C] lower than in neighboring areas. Any turbulence will cause the air to mix with the air above, and so frost-hollow effects are noticed only in calm anticyclonic conditions. One of the greatest frost hollows in the world is at Verkhoyansk in Siberia, Russia, where January average temperatures are about –49°F [–45°C]. In frost hollows, the coldest temperature is just above ground level; above that the air temperature is higher. Normally air temperatures decrease with height, and this converse situation in frost hollows

functions Collectively, the services, goods and amenities available in a settlement. A large city has many more functions than a small village. Low-order functions are those which provide **convenience goods** likely to be required frequently, perhaps every day, such as newspapers, bread and milk, and these will be found in most settlements. High-order functions, which supply specialized or long-lasting goods, will be available only in larger cities.

G

G8 (Group of Eight) The "inner group" of the **Organization for Economic Cooperation and Development (OECD)**, whose aims are to stimulate economic and social progress in its member countries. The G8 consists of Canada, France, Germany, Italy, Japan, Russia, the UK and USA, whose heads of government often meet to discuss major problems, such as recessions.

gabbro A coarse-grained **igneous** rock. The mineral content of gabbro consists mainly of **feldspars** and ferromagnesian minerals; for this reason it is darkish in color and alkaline in content. The coarse grains indicate that it was formed at depth beneath the Earth's surface, and was able to cool slowly, allowing time for the formation of large crystals. The equivalent acidic igneous rock with large crystals is **granite**.

gap An opening in a ridge, often created by a river. Some gaps are formed by glacial erosion, or by the effects of **meltwater** at the end of a glacial phase. Gaps often became important route ways; for example, Cumberland Gap in northeast Tennessee, USA, which was followed by many

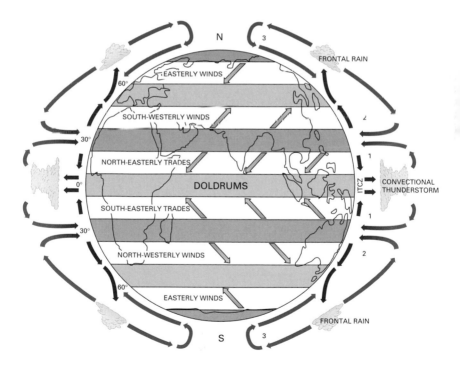

▲ **general circulation** Hot air rises at the Equator, at the intertropical convergence zone (ITCZ), causing an area of low pressure. The air cools and falls, creating the Hadley cells (1) of circulating air. The Hadley cells drive the Ferrel cells (2) and polar cells (3), which are also affected by temperature differences between places nearer to and further from the poles. Air circulation has a major influence on rainfall.

early pioneers and explorers from 1750 onward. The ease of passage through gaps often resulted in the development of settlements nearby. Many villages and towns can be described as gap towns.

garden city British name for a planned, self-contained settlement, first designed by Ebenezer Howard in 1898; it would have both residential and industrial areas, but would be rural in layout. It would thereby offer the benefits of urban living (such as recreational facilities) without the congestion that was to be found in 19th-century cities at that time. Garden cities would offer low-density housing, many parks and open spaces, and a maximum population size of about 30,000. The first city built in this style was Letchworth, England, in 1903.

garden suburb A suburb with a layout inspired by ideas for **garden cities**, with low-density housing and open suburbs. Such suburbs were built in the late 19th and early 20th centuries in Britain; for example, Hampstead Garden Suburb in 1907.

GATT SEE **General Agreement on Tariffs and Trade**

GDP SEE **gross domestic product**

General Agreement on Tariffs and Trade (GATT) A treaty, operating as a specialized agency of the UN, signed in 1947 by 23 countries (this had risen to 125 in 1994) which aimed to abolish trade barriers in international trade and expand world trade through setting up multilateral trade agreements. New proposals were developed periodically at "rounds" of talks; the Uruguay Round ended on December 15, 1993 after seven years of talks, with 117 countries signing 40 separate agreements. On January 1, 1995, GATT was replaced by a multilateral trade organization, known as the **World Trade Organization**.

general circulation The general movement of the **atmosphere**, giving particular pressure systems and winds seasonally or all year around. These winds take heat from the tropical latitudes and carry it to the poles to maintain the pattern of world temperatures.

gentrification The improvement of houses and the creation of new and often expensive houses in some **inner city** areas which have suffered from industrial decline. An example is the Upper West Side of Manhattan in New York City, USA. In such areas there is often a shortage of cheaper housing to suit the needs of the local people. However, in such areas there can be a shortage of buyers, leading to a different sort of decline.

geographical inertia The tendency of a place to continue with a particular industry or business activity after the original advantages afforded to that activity have disappeared. This is because the factory, machinery, local labor supply, tradition and reputation may allow the industry to continue successfully. Such is the case with the pottery industry in Stoke-on-Trent, England. Originally, the industry was established there because there were local supplies of clay and coal; nowadays, the clay is brought in from other areas, but the industry still continues. ALSO CALLED **geographical momentum, industrial inertia**

geographical momentum SEE **geographical inertia**

geography The study of the Earth's surface and its people. The two main subdivisions are physical and human. Physical geography includes **geomorphology, climatology, meteorology** and **pedology**; and human geography includes the study of agriculture, industry, resources and **political geography**. Geography also involves the study of such varied present-day topics as **conservation**, racism and traffic congestion, and is closely linked with many relevant issues of modern life. SEE ALSO **human geography, physical geography**

geological column A diagram which shows the **geological time scale**. The dates for the geological periods are based on radiometric dating, but should not be regarded as absolutely precise. The older periods, especially, may vary by a few million years. ALSO CALLED **stratigraphical column**

geological time scale The sequence and approximate duration of the geological periods. SEE ALSO **geological column**

geology The study of the origin and structure of the Earth and of the changes it has undergone and is in the process of undergoing. Geologists work from the crust inward, whereas geographers look at what is on the surface.

geometric rate A sequence of numbers in which the ratio of any number to its predecessor is a constant; for example, 2, 4, 8, 16, 32, 64, 128, etc. **Thomas Malthus** believed that world population would increase at a geometric rate, while the world's food supplies would increase at only an arithmetic rate. SEE ALSO **arithmetic growth**

◀ **geological time**
Time, in millions of years before the present, is shown on a sliding scale, which has been greatly compressed in the distant past. Geologists can date rocks accurately by radioactive dating, and particularly by using mass spectrometers to measure isotopes precisely.

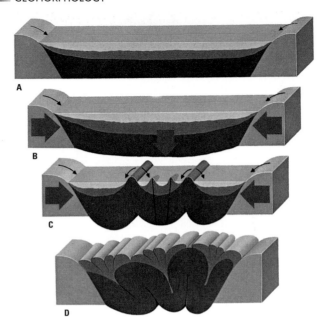

◀ **geosyncline**
Geosynclines are the birthplaces of mountains. They are large troughs where thick layers of sediments can accumulate (A). Where geosynclines develop between two colliding crustal plates (B) the sediments can be squeezed up as broad ridges known as geanticlines (C). Further compression creates mountain ranges (D). The whole process is usually accompanied by pressure-induced recrystallization or melting, which then forms metamorphic, plutonic and volcanic rocks; examples of each of these are gneisses, granites and rhyolites.

geomorphology The study of the shape of the Earth. Geomorphology considers the relationships between geological structures and surface landscape features, as well as the processes which change the features by **erosion** and **deposition**. Rivers, coastlines, rock types, slope formation, ice, erosion and weathering are all within the scope of the geomorphologist. ALSO CALLED **physiography**

geopolitics The study of nation states in the context of their geographical location and environment, and how this affects their development.

geosyncline A large **syncline** which may extend for hundreds of miles. It is a basin-shaped area located between two of the Earth's plates. As the plates move closer together, the geosyncline is compressed. Sediments which have accumulated on the floor of the geosyncline will be crumpled and buckled to form fold mountains. The Tethys Sea, an enlarged version of the Mediterranean that existed some 200 mil-

lion years ago when the continents first began to break up, was a good example of a geosyncline; as it was reduced in size, the Alps and Atlas Mountains were formed.

geothermal Relating to the heat of the Earth's interior. Geothermal activity produces geysers, springs, mud volcanoes and steam. Some of these phenomena have now been harnessed to provide energy. Iceland, Italy, New Zealand, and California, USA, are important producers of power of this kind. In Iceland, cheap heating is provided to the capital, Reykjavík, using hot water from underground. There are also many **greenhouses** which are warmed by underground heat. Geothermal heat is a clean form of energy and it is renewable.

geyser A natural spring which intermittently throws out jets of water and steam. Geothermal heat within the rocks, generally in areas which have had volcanic activity, warms up the **ground water**. Water which accumulates in a tube or vent can warm up to such an extent that the

GEOTHERMAL ENERGY

An artificial hot spring can be used as a source of heat energy. A borehole is drilled several hundred metres into a natural cavity in the Earth, in which the temperature may be as high as about 600°F (300°C). Water pumped down the bore is heated, turns to steam and is forced up a second borehole. At the surface the steam drives turbines to produce electricity.

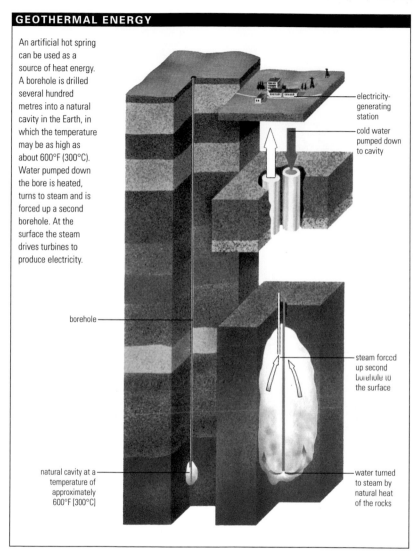

electricity-generating station

cold water pumped down to cavity

borehole

steam forced up second borehole to the surface

natural cavity at a temperature of approximately 600°F (300°C)

water turned to steam by natural heat of the rocks

water near the bottom of the vent begins to boil, causing the water higher up to blow into the air as a natural fountain. The erupted water falls back to earth, trickles back into the vent, and the warming process begins again. The name "geyser" is derived from the Icelandic *Geysir*, and the "Great *Geysir*" used to erupt every 30 minutes or so. It is now nearly extinct, but can be made to erupt, for the benefit of tourists, by feeding it with special additives. There are numerous other geysers in Iceland which do erupt naturally. The best known geyser in the USA is Old Faithful in Yellowstone National Park, Wyoming, which used to erupt every 67 minutes, but it, too, has become less reliable and quite irregular.

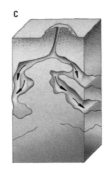

▲ **geyser** A plume of hot water and steam, a geyser is the result of the boiling of water at depth in a series of interconnecting chambers by volcanic heat (A). The expansion of steam produced drives the water and steam above it out at the surface (B), and this is followed by a period of refilling and heating making it a periodic phenomenon (C).

gher A circular tent covered in felt or skins used by nomadic peoples in Siberia, Russia, and Mongolia. ALSO CALLED **yurt**

ghetto An area in a city (often a slum) occupied by a minority or **ethnic group**. This term was first used to describe the enforced residential area of Jews in medieval Europe (for example, in Poland and Italy); its use has now spread to include other groups, for example, black minorities in the USA.

glacial Relating to a glacier. "Glacial advance" is a period of increasing snow and ice, when glaciers move down valleys. "Glacial retreat" is the reverse process. Less snow and ice means that the melt at the **snout** of the glacier is more than the replacement rate from the snowfield at the top of the glacier. As the glacier shrinks, it appears to be retreating. Glacial conditions are those experienced during a glacial phase, when subfreezing temperatures continue for most of the year. Much of Greenland and most of Antarctica are experiencing glacial conditions at present. SEE ALSO **ice age**

glacial deposition The laying down of material which has been transported by a glacier or ice sheet. The glacial deposits are left stranded on the landscape, and may be grouped together under the general term "**drift**." The commonest types of drift are known as boulder clay or **till**.

glacial erosion The process by which the Earth's surface is worn away by glacier ice. The presence of rock fragments within the ice and the action of meltwater streams accelerate the process.

glacial phase SEE **ice age**

glacier A mass of ice which flows down a valley. Snow accumulates on mountainsides, above the snow line, but gradually starts to move downhill because of the effects of gravity. As the snow is compressed by the weight of more snow on top of it, it is likely to melt and then refreeze into granular ice called **firn**. Continued pressure gradually turns the firn into ice. As it moves downhill, a glacier normally follows the route of an existing river valley, which is likely to be deepened, flattened, widened and steepened by the action of the ice. The glacier will grow and extend further down the valley, as long as the supplies of snow and ice from the snowfield are greater than the rate of melt at the lower end of the glacier. If a glacier continues to grow, it is said to be "advancing," but if it shrinks it is said to be "retreating." There is a seasonal advance in winter, and a retreat in summer, whatever the long-term changes. During the glacial periods, many glaciers extended down the valleys and on to the lowlands beyond. The ice spreads out on the plains to form ice sheets.

glasshouse SEE **greenhouse**

gley A type of soil which occurs in wet areas. Where the water table is near the surface and drainage is very poor, waterlogging will be the result. The soils will be greyish in color and probably contain some brownish deposits of ferric hydroxide. Gley soils are found in **tundra** areas and in wet, badly drained hollows in more **temperate** zones. ALSO CALLED **glei**

global warming The slow increase in world temperatures caused by the **greenhouse effect**. This is an example of **climatic change**.

GMT SEE **Greenwich Mean Time**

gneiss A metamorphic rock with a distinctive layering or banding. Because of heating during metamorphism, the different minerals melted and then solidified in

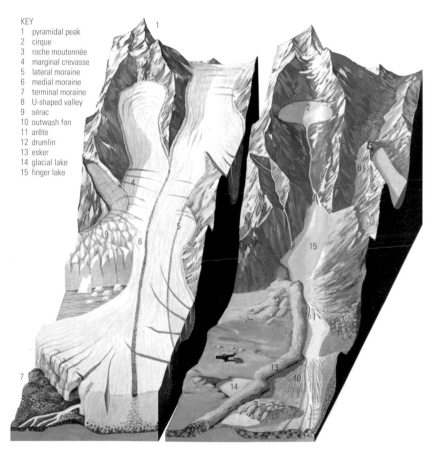

KEY
1 pyramidal peak
2 cirque
3 roche moutonnée
4 marginal crevasse
5 lateral moraine
6 medial moraine
7 terminal moraine
8 U-shaped valley
9 sérac
10 outwash fan
11 arête
12 drumlin
13 esker
14 glacial lake
15 finger lake

▲ **glacier** In spite of a return to warmer conditions, some regions of the world are still covered by ice, and are being greatly altered by its action. Glaciated regions have been subjected to erosion and deposition, the erosion mainly taking place in the highland areas, leaving features such as pyramidal peaks, cirques, roches moutonnées, truncated spurs and hanging valleys. Most deposition has occurred on lowlands, where, after the retreat of the ice, moraines, drumlins, eskers, erratic boulders and alluvial fans remain.

GLOBAL POSITIONING SYSTEM (GPS)

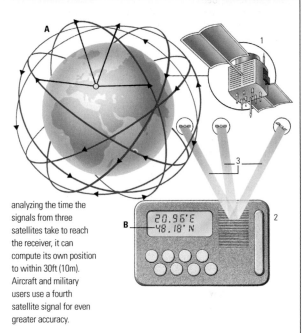

The world is ringed by 24 global positioning satellites (1) launched by the USA. At any time four are above the horizon wherever you are on Earth (A). With a handheld receiver (2) that compares the time signals (3) from the satellite, your position can be fixed to a remarkable degree of accuracy. The receivers are now small enough to fit in the hand, and display either longitude and latitude (B) or a grid reference. Each satellite carries an atomic clock and transmits time signals to Earth. The receiver knows exactly where each satellite should be at any given time, and by simultaneously analyzing the time the signals from three satellites take to reach the receiver, it can compute its own position to within 30ft (10m). Aircraft and military users use a fourth satellite signal for even greater accuracy.

---

distinct layers. The darker minerals are likely to be hornblende, augite, **mica** or dark **feldspar**. Before metamorphism, gneiss was an **igneous** rock, possibly a **granite**. Good examples of gneiss can be found in northwest Scotland, especially on the island of Lewis. SEE ALSO **metamorphic rocks**

GNP SEE **gross national product**

Gondwanaland SEE **continental drift**

gorge A narrow steep-sided valley. Gorges are found in hard rock, otherwise the steep sides would be rapidly eroded. **Canyons** are examples of large gorges, and the largest examples of these can be seen in the southwest USA, such as the Grand Canyon in Arizona.

government incentives Offers made by governments to attract the development of industry in particular areas. Examples include grants for buildings and machinery, free rent of a government-owned factory, tax relief on investments or loans, etc. It is an attempt by a government to stimulate development in a poor area and to decentralize industries from large urban locations.

graben SEE **rift valley**

graded profile A smooth profile of the course of a river. It occurs when all irregularities such as waterfalls have been eliminated. The longitudinal profile of a river valley is a smooth curve, whereas the **long profile** of a glacial valley will have irregularities.

gradient I A slope which can be shown on contour maps by the distance between the **contours**. The closer the contours are, the steeper is the gradient. In order to calculate the average gradient between two points, it is necessary to work out the difference in height between the points, and to measure the distance. The difference in altitude divided by the distance apart will give the gradient; for example,

$$\frac{500}{5000} = \frac{1}{10}$$

This gives a gradient of one in ten (1:10).
2 A curve on a graph used to represent a rate of change in relation to distance, as, for example, the decline of population density from the center of a city outward.

granite A coarse-grained acid **igneous** rock. It consists of large grains visible to the naked eye, which means that it was formed beneath the Earth's surface and cooled slowly. This enabled the liquid **magma** to solidify into large crystals. Granite is light in color because of its high acid content, and its main minerals are **quartz**, **feldspar** and **mica**. Granite is a hard rock and slow to weather. Although formed underground, it is often seen on the surface because all the other overlying rocks have been eroded. For this reason, too, it generally forms upland regions. Weathered granite disintegrates, leaving kaolin as a residual mineral.

graph A diagram illustrating a relationship between two variables, usually with two axes at right angles to each other. Each variable is then scaled along one of the axes.

grassland A region in which the natural vegetation is mainly grass. Grasslands occur in areas where the rainfall is too heavy for a desert but insufficient to support a forest. The total annual rainfall is likely to be more than 10 in [250 mm], and even as much as 40 in [1,000 mm]. The seasonal distribution may be important, because in some tropical areas where evaporation is very high, the effectiveness of summer rainfall is very low. For this reason, some locations with 40 in [1,000 mm] rainfall are unable to support woodland. There are numerous varieties of grass, but two main types of grassland; these are the temperate and the tropical grass-

lands. Temperate grasslands are found in central North America, where they are called **prairies**, and in Russia, where they are called **steppes**. An extension of steppe conditions can be found in Hungary. There are also temperate grasslands in the southern hemisphere: on the **pampas** of Argentina, the **veld** of South Africa, the Murray-Darling basin in Australia, and the Canterbury Plains of South Island, New Zealand. The steppes and the prairies have a continental type of climate, with warm summers and summer rainfall, but dry and very cold winters. Temperature ranges from 60°F to 70°F [15°C to 20°C] in July, and 15°F to −5°F [−10°C to −20°C] in January, with a total precipitation of 20 in [500 mm]. The southern hemisphere grasslands are generally monsoonal and are located in narrower land masses. Winters are milder, 40°F to 50°F [5°C to 10°C]; summers are about 70°F [20°C], and the rainfall is heavier (20 in to 30 in [500 mm to 750 mm]) and is less concentrated in the summer months. The grass in the temperate areas is fairly short, remains greenish throughout the year, and provides good pasture for animals, although much of it has now been plowed up for cereals. **Tropical grasslands**, or **savannas**, have very tall grass, up to 6 ft [2 m], in the hot wet summers, but in the hot dry winters the grass goes brown and shrivels. The total rainfall may be 40 in [1,000 mm], but most falls in summer, when evaporation is high. Summer temperatures are 80°F [25°C] or more, and winters are about 60°F to 70°F [15°C to 20°C]. Tropical grasslands are found in East Africa, Nigeria, northern Australia, the Orinoco basin in northern South America, and parts of the Brazilian plateau.

grass minimum thermometer A thermometer used to record the minimum temperature 1 to 2 in [20 mm to 30 mm] above ground level, where the air temperature often reaches its lowest point. A metal rod reveals the lowest temperature experienced during the night.

gravel A coarse **sedimentary rock**, often unconsolidated. Gravels may be cemented, to form **conglomerate** if the particles are rounded, or **breccia** if the fragments are angular. Unconsolidated deposits may be found on river terraces or in areas of glacial drift.

gravity The pull force exerted by one body on another. The force decreases with distance. For example, the gravitational force of the Earth means a body on its surface or in its gravitational field is kept on the Earth's surface.

gravity model A model which analyses or studies some kind of movement. It is generally used for investigating and attempting to predict the movements between two particular locations. It is thought likely that the population movements between two settlements will be proportional to their size; also, that the attractions of two settlements to the population of a third settlement nearby will be proportional both to the size of the two and also to the distances at which they are situated. SEE ALSO **Reilly's Law of Retail Gravitation**

great circle An imaginary circle drawn around the Earth. If a cut was made through the center of the Earth along the line of a great circle, it would split the Earth in half. The **Equator** and all lines of longitude are great circles, and so too are numerous diagonal lines. The shortest distance between any two points on the surface of the Earth will follow the line of a great circle. Sailing vessels used to follow great circle routes, and nowadays many air routes, especially over the Arctic, follow the line of a great circle.

green belt An area of green, mainly rural, land surrounding an urban area. The major green belt in the UK is the metropolitan green belt, which surrounds London, but there are other green belts around other major cities. There are strict controls over

GRAVIMETRIC ANALYSIS

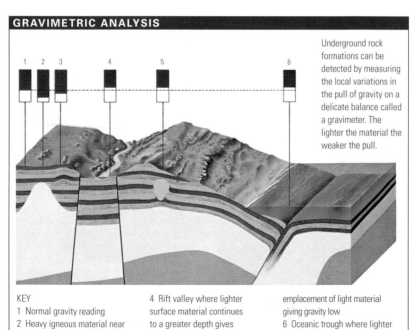

Underground rock formations can be detected by measuring the local variations in the pull of gravity on a delicate balance called a gravimeter. The lighter the material the weaker the pull.

KEY
1 Normal gravity reading
2 Heavy igneous material near surface gives high reading
3 Anticline gives gravity high
4 Rift valley where lighter surface material continues to a greater depth gives gravity low
5 Salt dome or upward emplacement of light material giving gravity low
6 Oceanic trough where lighter crustal material deep in mantle gives gravity low

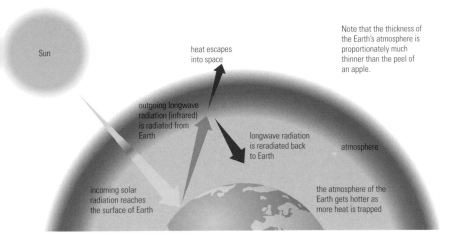

Sun

heat escapes
into space

Note that the thickness of
the Earth's atmosphere is
proportionately much
thinner than the peel of
an apple.

outgoing longwave
radiation (infrared)
is radiated from
Earth

longwave radiation
is reradiated back
to Earth

atmosphere

incoming solar
radiation reaches
the surface of Earth

the atmosphere of the
Earth gets hotter as
more heat is trapped

▲ **greenhouse effect** Energy from the Sun enters Earth's atmosphere, where some is absorbed. The rest warms the surface of the Earth, from which it is reradiated as longwave infrared radiation. Some of this radiation is absorbed in the atmosphere by greenhouse gases, such as carbon dioxide and methane. Some of the absorbed radiation is reradiated out into space, but some is reradiated back to Earth. If there is an increase in greenhouse gases, the amount of heat returning to Earth will increase.

building and development in green belt areas, although agricultural buildings may be constructed. Planning permission is necessary before any kind of urban development takes place, and urban sprawl is prevented from spreading into green-belt areas.

greenhouse A building, mainly of plastic or glass, for the cultivation of plants. Greenhouses are used to create an artificial climatic environment in order to protect or grow crops in the winter, or produce earlier vegetables than is possible out of doors. Some greenhouses are heated to enable growth throughout the year.

greenhouse effect The warming of the lower layers of the **atmosphere** as a result of retention and reradiation of solar heat. As the Sun's rays pass through the atmosphere on their way down to Earth, some heat is absorbed, but much of the solar energy passes through and is absorbed by the Earth. This energy is given off by the Earth as long-wave (infrared, or heat) radiation, which cannot pass easily through the atmosphere and is partially absorbed. Part

of this absorbed heat is reradiated to space, but some is reradiated back to Earth. The atmosphere thus acts rather like the glass in a **greenhouse** and traps heat. This effect, which helps to maintain the Earth's temperature, is called the "greenhouse effect." Certain gases are a greater barrier to the longwave radiation than others, and these are known as "greenhouse gases." Important examples are carbon dioxide and methane. Human activity, particularly burning **fossil fuels**, which produces millions of tons of carbon dioxide every year, is increasing the amount of the greenhouse gases in the atmosphere. This may lead to climatic conditions warming up slightly. If this process continued for the next few decades, it would cause a considerable melt of Antarctic and Greenland ice, which would cause a rise in sea level and the flooding of several of the world's major cities, including New York, USA and London, England.

Greenpeace A conservation group, which is active in many parts of the world. It will support any kind of protest its members consider relevant to their ideas of protecting and preserving wildlife. As a group they are opposed to the construction and use of nuclear power stations and nuclear weapons.

green revolution A period when rapid progress was made in the production of extra food supplies. In the 1960s, research at the Rockefeller Center in Mexico produced new strains of wheat and corn which could produce higher yields per acre. At about the same time, research in the International Rice Research Institute in the Philippines produced several new **rice** hybrids. Suddenly the world was able to produce more food. The development of new hybrid varieties is a continuing process; at the same time, new pests evolve, and so it is a constant race to keep ahead of all the problems. SEE ALSO **high-yield variety, hybridization**

Greenwich Mean Time (GMT) The local time at longitude 0°, which passes through Greenwich, in London, England. It is the **standard time** for Britain, from which most other world times are calculated.

grid A pattern of vertical and horizontal lines drawn on a map in order to provide a basis for map references. The numbering of a grid (vertical and horizontal) starts in the bottom left-hand corner of the map. The vertical grid lines are used to determine the easting (that is, the location of a point to the east of the nearest grid line); while the horizontal lines determine the northing (that is, the location of a point to the north of the nearest grid line).

grid-iron drainage A drainage system in which there are a number of parallel streams, each with tributaries that run at right angles to the main stream, so that there exists a rectangular network of streams. Examples can be seen in the Appalachians in the USA, and the Jura Mountains in Europe. ALSO CALLED **trellis-work drainage**

grid-iron pattern A pattern of roads shaped like a "grid iron." The roads run parallel to each other, with a second set that cross at right angles. This rectangular plan which is common in many towns in the USA, is often the result of careful planning before much building has taken place. Most European towns have not been planned, and therefore have a more irregular plan, although the central parts of some towns and some sections of new towns show a grid-iron pattern.

grike An enlarged joint in **Carboniferous** limestone. Where **solution** has weathered away at a **joint**, a vertical crack is opened. Grikes may be 3 feet [1 m] or more in depth and there are generally several of them in close proximity. Down in the sheltered hollows created by the grikes, there is often a wealth of small lime-loving plants. ALSO CALLED **gryke**. SEE ALSO **limestone pavement**

grit A coarse type of sandstone in which the grain size is larger than in a typical sandstone. A good example of grit is the **millstone grit** on which the city of Bowling Green, Ohio, in the USA is built. Millstone grit is also found in many parts of the Pennines in England. It gives rise to bleak moorland, with large expanses of **peat** bog, and there are often small but steep rocky outcrops called "edges."

groine A low barrier of concrete or wood built out into the sea from a seawall or promenade. It slows down the movement of **longshore drift**, and also protects the seawall from the direct battering of the waves. The length and frequency of groines depends on the angle at which the prevailing waves come in to the shore. Many coastal towns erect groines in order to protect the coast and save money on sea defenses, since groines are more cheaply repaired and replaced than buildings or roads. ALSO CALLED **groyne**

gross domestic product (GDP) The total value of goods and services produced by a country over any particular period, normally a year. All sections of the economy contribute to the GDP. If the total gross domestic product is divided by the total number of the population, a GDP per capita figure can be derived. This is a useful general guide to the wealth and standard of living of a country.

gross national product (GNP) The

value of the gross national product of a country is made up of the gross domestic product plus the money earned by investments abroad, but minus the money earned by foreigners living in the country.

ground frost SEE frost

ground water Water which is found

underground, in the pores and cracks of the rock. It consists mainly of rainwater, which has percolated down into the rocks. Most ground water is near the surface, and if it reaches the surface it will cause a puddle or a lake. Ground water moves downward because of gravity; it may move sideways through rocks, especially where an impervious stratum prevents downward

movement. In some rocks there may be considerable movement of water, and this is particularly marked in **Carboniferous** limestone. Underground streams can sometimes be seen, which have created many miles of tunnels and caverns. As it moves, the water can cause erosion and chemical solution. A wide variety of chemicals may be dissolved from the rocks by the effects of ground water.

Group of Eight SEE G8

growing season The period in which plants normally grow. Only grass can survive on hills and so there is usually **pastoral farming**, but not **arable**. In central USA, the "Corn Belt" is commensurate with the area which has 150 frost-free days.

▼ **groine** Groines are one of a number of common sea defences designed to stop the erosion of the shoreline. The aim of the defences is to absorb the force of large waves and to resist gradual erosion by smaller waves and currents.

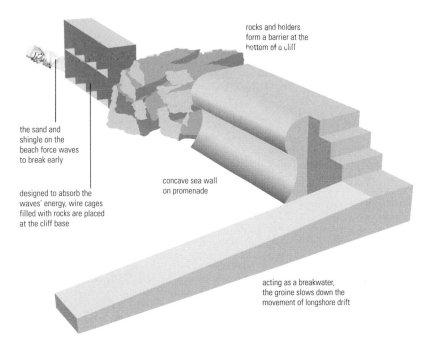

rocks and bolders form a barrier at the bottom of a cliff

the sand and shingle on the beach force waves to break early

concave sea wall on promenade

designed to absorb the waves' energy, wire cages filled with rocks are placed at the cliff base

acting as a breakwater, the groine slows down the movement of longshore drift

If the growing season is shorter, wheat is the major crop; if the growing season exceeds 200 frost-free days, cotton is grown.

growth pole A location in which industrial development and economic expansion is taking place. This tends to be an **urban** area, although in advanced countries it is possible to set up industries in a wide range of locations, because transportation, water and electricity (**infrastructure**) are widely available. In poorer countries, there are few locations where developments are possible because of the lack of infrastructure, and so most industries congregate in the capital, main port, or one of the few large towns. In order to help developments in less favored areas, it is important to create new centers of growth, or "growth poles." In Brazil, the development of the new capital, Brasília, created a new growth pole. In Venezuela, the developments in the Orinoco and Caroni valleys near Ciudad Guayana are further examples of growth poles.

groyne SEE **groine**

gryke SEE **grike**

Gulf Stream An ocean current which flows out of the Gulf of Mexico, and northward along the east coast of the USA. As it move northward, it begins to turn to the right because of the effects of the rotation of the Earth. It then begins to cross the Atlantic Ocean; at this point, it is more correctly known as the North Atlantic Drift. As it originates in the Gulf of Mexico, which is one of the warmest parts of the oceans, it is a very warm current, about 80°F [25°C]. While crossing the Atlantic it cools down gradually, but it still brings much warmth across to the eastern side of the Atlantic. Some of the North Atlantic Drift flows along the Norwegian coast, which is kept ice-free as far east as Murmansk in Russia. The current is a great source of winter warmth for the UK and Ireland, and helps to create winters which are very mild for the latitude. The Isles of Scilly off Cornwall and the southwest tip of Ireland are able to grow various subtropical plants because of the January average temperature of 45°F [7°C], which is warmer than along the Mediterranean coast of France at that time of year. SEE ALSO **ocean currents**

gully A narrow channel on a hillside or slope, formed by the action of water, often because of rapid runoff following heavy rain. Gully erosion, often referred to as "gullying," is the result of rapid runoff on a slope which has no vegetation cover. In southern Italy and other Mediterranean lands there are many badly gullied areas, where the vegetation has been removed by man and domestic animals, so that the soil has been washed away by heavy rain. There are many areas with serious gullying in southern Africa and in New Zealand, and most notably in the Badlands of South Dakota in the USA.

H

habitat An area in which an organism can live, with relatively favorable environmental conditions.

hachuring SEE **relief map**

hacienda A large estate, especially in the Spanish-speaking countries of South America. Many haciendas have absentee landowners, who leave a manager in charge. They often exploit Indian and **mestizo** labor. Most haciendas grow a cash product, which may be wheat, cattle or grapes, but generally they also grow food for the workers.

hade The angle of declination of a fault plane from the vertical.

Hadley cell Atmospheric circulation cell, named after the British scientist George Hadley. Hadley proposed it (1735) to explain the trade winds, in which winds rise and flow polewards from the Equator and then descend and flow Equatorwards, transferring heat by convection. SEE ALSO **general circulation**

hail Small pellets of ice which have been formed by raindrops freezing. When convective currents rise quickly and form large **cumulonimbus** clouds, the water vapor in these clouds may be carried up to great heights. Large drops of rain may be formed, but if the uplift continues, drops may freeze. As they become too heavy to be carried up further, they start to fall, gathering more water as they descend. A further burst of rising air, containing convective currents which are pulsating rather than continuous, will force them to rise again. The extra water may be frozen to form a bigger hailstone. The greater the number of times a hailstone has to rise, the bigger it will become. In temperate latitudes, hailstones rarely exceed pea size, but in the tropics, where there is a greater amount of convection, stones the size of golf balls some-times occur. These may weigh up to 1 lb [0.5 kg], and can do considerable damage when they land.

hanging valley A valley which is elevated above the level of a main valley. The main valley will be a **U-shaped valley** which has been deepened by a **glacier**. The erosion rate in the tributary valley having been slower than in the main valley, it was not as deeply cut. The tributary valley may be U-shaped if it contained a tributary glacier, or it may be a **V-shaped valley** cut by a river. The stream will flow down steeply from the hanging valley, into the main valley, and it is quite likely that there will be rapids or a waterfall on the tributary stream. Some hanging valleys in Lauterbrunnen in Switzerland and Yosemite in California, USA, drop almost vertically for more than 1000 ft [300 m]. In England there are impressive hanging valleys in the Langdale and other Lake District valleys.

harbor A stretch of water close to the shore which provides an anchorage and shelter for boats and ships. It is usually protected by natural or man-made walls, and boats enter through a narrow harbor mouth.

hardpan A layer of hard material found just below the surface, often in sandstone areas. Rainwater leaches away chemicals including iron oxide and calcium carbonate, and sometimes these solidify 3 ft to 6 ft [1 m to 2 m] below the surface. If this happens, they form an **impervious** layer – hardpan – above which there may be marshy patches or puddles, even on **permeable** rocks. If the hardpan layer is mainly iron oxide, it is sometimes referred to as "iron pan."

hard water Water which contains a high concentration of dissolved carbonates of calcium and magnesium, derived from the rocks through and over which the water has flowed; for example, limestone. Hard water does not easily form a lather with soaps or detergents.

hardwood A broad-leaved, as opposed to a coniferous, tree. Hardwood trees are generally much slower growing than conifers and may take 80 to 100 years to reach maturity. Hardwoods in North America and northern Europe include oak, maple, walnut and cherry. Evergreen oaks and eucalyptus are important hardwoods in the Mediterranean countries and Australia. Monsoon lands and tropical forests contain teak, mahogany and rosewood.

harmattan A strong northerly or north-easterly wind which blows from the Sahara Desert to Nigeria, Ghana and other countries on the Guinea coast of West Africa. It is a hot dry wind. If it blows over the humid coastal areas it seems cool and refreshing, and is sometimes called "the doctor." However, it can carry dust from the desert and may be so dry that it shrivels the crops. It is likely to occur throughout the year between 10°N and 15°N, but will only reach south to the coast in the winter months.

Harris and Ullman model One of the classical models of urban morphology. SEE ALSO **Ullman, E. R., and Harris, C. D.**

headward erosion A type of erosion which occurs at the source of a stream. It lowers the land slightly, allowing the stream to rise further back, or further uphill. Over long periods of time, it is possible for streams to erode through **watersheds**, and this enables them to capture streams flowing in other valleys.

heat island SEE **urban heat island**

heave SEE **fault**

heavy industry Any of the industries which produce or use large quantities of bulky raw materials, such as coal-mining, steel-making, shipbuilding, and various chemical and engineering concerns. Most of these industries first developed in coal-mining regions, but many are now located near ports.

hectare A unit of area, equal to 2.47 acres. One hectare equals 10,000 sq m [107,640 sq ft], while 100 hectares are 0.4 sq miles [1 sq km].

hematite A type of iron ore. It is a rich deposit, often as much as 75% in purity, but always at least 60% pure. It is frequently found in small lumps, which are referred to as "kidney ore" because of their shape. Many large deposits have been found; for example, at Kiruna in Sweden, in Western Australia, and near Lake Superior in the USA.

herbivore A plant-eating animal.

Hercynian A period of major mountain formation which occurred at the end of the **Carboniferous** period and the beginning of the Permian, about 250 million years ago. The name "Hercynian" is derived from the Harz Mountains in Germany, and the alternative name "Armorican" is named after Brittany, which was called "Armorica" in Roman times. ALSO CALLED **Armorican**

heritage coast Canadian and British term for a coastline that has been unaltered and protected from development because of its natural beauty and wildlife. The equivalent in the USA is "national seashore". Canada's Great Lakes Heritage Coast extends 2,604 m (4,200 km).

hierarchy of settlements A system of grading the various types of human settlement according to size, devised by Walter Christaller in his **central place theory**.

high SEE **anticyclone**

high technology The use of highly developed and sophisticated equipment and machinery. It normally involves spending huge sums of money on equipment, but utilizing small amounts of labor. The labor is likely to be highly skilled, trained and educated, and highly paid. Examples of high technology include work with electronics, computers and silicon chips, but also new machinery, such as may be found in very modern automated car factories or steel works.

high-yield variety A crop that has been bred to produce more than it would in its natural form. Such crops often need large quantities of fertilizers and pesticides to ensure they do well, but have been of great benefit in reducing food shortages in developing countries. It has also been possible to produce species that are resistant to drought and disease. One example of a high-yield variety is IR8, a new variety of rice, developed in the Philippines, that has produced up to six times more yields. SEE ALSO **green revolution**, **hybridization**

hill farming A system of farming which is generally extensive and pastoral. Sheep and sometimes beef cattle are allowed to roam on the hills and fend for themselves. They have to be rounded up at certain times of the year; for example, for dipping, lambing or selling. The farms are normally located in the valleys, and the animals are often kept near the farms during the winter. A few crops such as hay and oats may be grown as fodder. As soon as the weather improves in the spring, the animals are moved up the slopes.

hill fog SEE **fog**

hinterland 1 The sphere of influence of a port. It is the area from which exports are obtained, and the area which receives goods that have been imported through the port. The hinterland of a large port, such as London in England, may cover the entire country, but smaller ports, such as Savannah, Georgia, and Newport News, Virginia, in the USA, have a more localized hinterland. The boundaries of hinterlands often overlap, since some goods could pass through any of a number of ports. A hinterland may also be international; for example, Rotterdam handles goods for Germany, Switzerland, France and Belgium, as well as the Netherlands. 2 The

area served by a large inland city, not necessarily a port.

histogram A graph which shows groups of classes of data by means of a series of columns or rectangles placed side by side. The intervals of values are placed along the horizontal axis, and the height of the column above each value band shows the corresponding frequency. It is a very clear way of summarizing information.

hoar frost SEE **frost**

hogsback A narrow steep-sided ridge which stands up sharply from the surrounding countryside. It may be the result of folding or faulting, followed by differential erosion, in which the softer neighboring rocks have been eroded more quickly.

Holocene The most recent period (or epoch) of geological time, stretching from 12,000 years ago to today. SEE ALSO **geological column**, **geological time scale**

horizon Any of the layers of soil which can be seen in a soil profile. The normal divisions are the A, B and C horizons; A is the **topsoil**, B the **subsoil**, and C the **bedrock**. The A horizon contains fine particles, humus and decaying vegetable matter. The B horizon contains some fine particles as well as coarser fragments, and contains much more inorganic material than the A. It also contains minerals washed down from the A horizon by leaching. The C horizon is rocky. The boundary lines between these zones are generally rather vague. It is becoming increasingly common to divide the C horizon into C and D. In this case, the C is weathered rock fragments, and D is the real bedrock. SEE ALSO **A horizon**, **B horizon**, **C horizon**

horn A pyramid-shaped peak such as the Matterhorn in the Alps. The pyramid would have been steepened and carved by glacial erosion, the ice action and **freeze-thaw** activity having formed **cirques**, which wore away the sides of the pyramid. The summit thus became smaller and steeper to form a peak. It is often possible to reach the summits of horns by ascending along the line of **arêtes** between the corries. Mt. Everest is a horn. ALSO CALLED **pyramidal peak**

horse latitudes Either of two areas of high pressure found near 25°N and 25°S of the Equator. The air rises at or near the Equator because of low pressure and then circles in the **atmosphere**, but much falls near the tropics of Cancer and Capricorn. The falling air gives rise to fairly permanent regions of high pressure, the "horse latitudes." Both the **trade winds**, which blow from the horse latitudes toward the Equator, and the westerly winds, which blow from the horse latitudes toward the poles, are caused by the high pressure. It is suggested that the horse latitudes were so called because in the days of sailing vessels it was easy to become becalmed in these areas, in which case everything possible would be thrown overboard in order to lighten the vessel. Apparently, it was not unknown for the horses to be discarded at this time.

horseshoe lake SEE **ox-bow lake**

horst SEE **block mountain**

horticulture The growing of vegetables, flowers and fruit. It is generally carried out on small farms or in market gardens, and is very intensive; there is much horticulture in **greenhouses**, too. Horticultural concerns are normally located near towns, in order to get the produce to market quickly; for example, just outside New York in the eastern USA. There are also some favored areas where early vegetables can be grown because of climatic advantages. For example, in the USA, Florida grows early fruit and vegetables, which are sent north to New York, and California also grows early products. ALSO CALLED **market gardening**, **truck farming** (*in the USA*)

hotspot An area on the Earth's surface where the crust is quite thin and volcanic activity can sometimes occur, even though it is not at a plate margin. The Hawaiian Islands are at a hotspot in the northern Pacific Ocean.

Hoyt, H. An American researcher who developed the **sector model** of urban growth. In 1939 he wrote about his views of urban growth, based on his studies of land-use patterns in 142 American cities. His ideas differed from those of Burgess, who regarded concentric growth as the most important, whereas Hoyt saw that sectoral variations could be related to the development of transportation links. His is one of the three classical models of **urban morphology**: the **Burgess model**, Hoyt's sector model, and the **Harris and Ullman model**

huerta An irrigated, highly cultivated market garden located along the eastern coast of Spain, where several crops of fruit and vegetables may be produced in a year. It is an example of **intensive farming**.

human geography A major section of geography which includes the study of people, work, lifestyles, and settlements. It also considers how man influences and changes the **environment**, as well as being influenced by it.

humidity The amount of water vapor in the **atmosphere**. SEE ALSO **absolute humidity**, **relative humidity**

humus The dark organic material in soils, produced from plant and animal matter. As plants and animals die and decay, their decomposing remains in the soil form humus. It will be plowed up in cultivated land, and possibly added together with manure and fertilizer. In areas where there is no cultivation, it gradually decays and is washed downward by rain. Some of it is then lost by **leaching**, but much will be reused by the next generation of plants. The humus layer varies from soil to soil. It is often black in some of the richer soils, but even in the best soils it can be used up by repeated plowing, especially if **monoculture** is practiced.

hunter and gatherer A member of a tribe or group of people, especially in areas of tropical forest, who survive by hunting animals and birds, and gathering wild fruits and berries. They will also fish to add to their food supplies. They have to move around from place to place in the search for food, and so they do not build permanent homes. Lean-to shelters of leaves and branches are often adequate in the warm climatic conditions of the tropics. The **pygmies** are an example of this way of life, and other groups are found in Africa, South America and Southeast Asia. Only small numbers of people still pursue this way of life.

hurricane **I** A tropical cyclone in the Caribbean or Central America. **2** A wind of hurricane force, i.e. force 12 on the **Beaufort scale**.

hybridization The crossbreeding of plants to produce new varieties, which are developed for special qualities, such as resistance to disease or pests, or the ability to ripen more quickly or give a greater yield. Hybrid varieties have been produced for many centuries, and even the earliest cultivated plants were selected from natural crossbreeding. Research stations, such as the International Rice Research Institute in the Philippines, continue to produce new varieties each year. SEE ALSO **green revolution, high-yield variety**

hydraulic action The erosive effect of water. The force of moving water can be an effective agent of **erosion**, rather like a battering ram. It is important both in rivers and on coastlines. In rivers the movement of currents causes hydraulic action, though if a load of silt or sand were added the process would be called **corrasion**. Along the coast, breaking waves sometimes trap water in cracks and joints, and this adds to the hydraulic effect.

HYDROELECTRICITY

A pumped storage hydroelectric system uses turbines to generate electricity at peak times and to pump water back behind a dam when demand is low. A reaction water turbine (1) drives an electrical generator (2). When the centrifugal pumps (3) are uncoupled the machine acts as a normal hydroelectric generator. But when the geared coupling (4) is engaged the water turbine drives the pump up to operating speed. Then the generator is connected to the electricity supply whence it acts as a motor. The turbine valve (5) is closed, the pump valve (6) is opened and water is pumped back behind the dam, adding to the volume of stored water available for later hydroelectric generation.

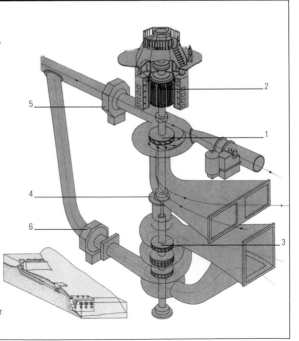

WORLD HYDROELECTRICITY

Percentage of electricity generated by hydro-electric power stations (leading nations, 1997)

	Country	Percentage
1	Paraguay	99.9%
2	Congo (D. Rep. of)	99.7%
3	Bhutan	99.6%
4	Zambia	99.5%
5	Norway	99.4%
6	Ghana	99.3%
7	Congo	99.3%
8	Uganda	99.1%
9	Burundi	98.3%
10	Uruguay	98.0%
11	Rwanda	97.6%
12	Malawi	97.6%
13	Cameroon	96.9%
14	Nepal	96.7%
15	Laos	95.3%
16	Albania	95.2%
17	Iceland	94.0%
18	Brazil	92.2%
19	Honduras	87.6%
20	Tanzania	87.1%

hydroelectric power Electricity produced from falling water. Natural or artificially created waterfalls provide the pressure of water needed to drive turbines to generate electricity. If there are no waterfalls, dams are built, and the water drops from near the top of the dam down to a power station at the foot. The water may also be used for irrigation. Most hydroelectric power stations are located in highland areas where the climate is quite wet. The world's leading producers of hydroelectric power are Canada and the USA. In recent years, because of improved technology, it has become possible to harness the power of large slow-flowing rivers, such as the Angara in Siberia, Russia, and the Amazon in Brazil. At some hydroelectric schemes, the electricity produced during the night, when there are not as many customers, is used to pump water back into the lake or reservoir, so that it can be used again to produce electricity.

hydrological cycle The water cycle. Water is evaporated from the sea; some falls back into the sea as rain, but much is carried over the land. There it falls as precipitation, and by **surface runoff** or **infiltration** and seepage it gradually moves back into the sea. Less than 1% of the world's water is involved in this cycle at any particular time; 97% of all water is in the sea, and much of the remainder is held as snow or ice.

hydrology The study of water on or under the Earth's surface, and its uses, availability, conservation, etc.

hydroponic The cultivation of plants in nutrient solutions rather than in soil. Soil provides plants with nutrients and a medium in which they can stand up. If those two requirements are met in other ways, soil becomes irrelevant. Nutrients can be provided in the water, and trays of gravel, sand, stones or even polystyrene make a suitable growing medium. Hydroponic techniques

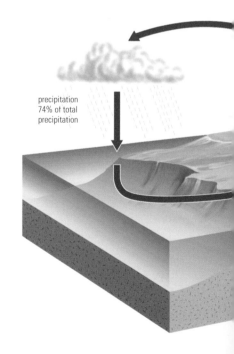

precipitation
74% of total
precipitation

are very useful in places where there is a shortage of soil, such as some of the semi-arid parts of Africa. Such areas have long hours of sunshine, which is very suitable for growing plants, providing they are watered and fed artificially. Hydroponic methods are also used in the Arctic areas of Canada and Russia, where long hours of sunlight in summer enable fresh vegetables to be grown. They are important in the production of food in several Middle East countries, and also in Singapore, where there is a great shortage of land.

hygrometer An instrument for recording the relative humidity of the air. It consists of two thermometers, one a dry bulb and the other a wet bulb. The wet bulb is

covered by a small piece of muslin cloth, which is kept damp by water from a small reservoir at the end of the hygrometer. Evaporation from the cloth reduces the temperature of the wet bulb, and the difference in temperature between the wet bulb and the dry bulb shows the relative humidity. Special tables are available to give the correct humidity. The two thermometers can be kept in a **Stevenson screen**, or can be placed on a mount, which can be whirled around. This type of hygrometer can be used out in the field to give approximate shade temperatures in an open and exposed location.

hypabyssal A term denoting **igneous** rocks which formed neither at the Earth's

▼ **hydrological cycle** The Earth's water balance is regulated by the constant recycling of water between the oceans, the atmosphere and the land. The movement of water between these "reservoirs" is called the hydrological cycle. The oceans play a vital role in this

cycle: 74% of the precipitation falls over the oceans and 84% of the evaporation comes from the oceans. Water vapor in the atmosphere circulates around the planet, transporting energy as well as water itself. When the vapor cools it falls as rain or other precipitation.

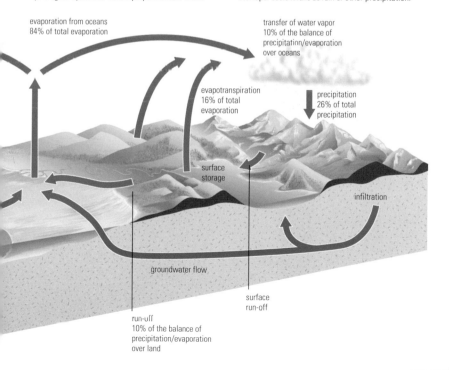

evaporation from oceans
84% of total evaporation

transfer of water vapor
10% of the balance of
precipitation/evaporation
over oceans

evapotranspiration
16% of total
evaporation

precipitation
26% of total
precipitation

surface
storage

infiltration

groundwater flow

surface
run-off

run-off
10% of the balance of
precipitation/evaporation
over land

surface, nor at great depth. They sometimes contain a few large crystals as well as many small crystals, and are intermediate in grain size between **plutonic rocks**, which were formed at depth, and **volcanic rocks**, which formed at the surface. **Hypabyssal** rocks are often found in **sills** and **dikes**.

hypermarket A very large store or shopping complex selling a wide range of necessities and consumer goods. Normally located on the outskirts of towns, hypermarkets include a large supermarket and various other types of store, as well as ample car parking space, possibly a garage, a children's play area and a bank. They developed first in the USA because of the wealth which enabled most families to become car owners. France was the first European country to develop similar out-of-town shopping complexes, but the idea has now reached Britain and many other countries.

hypolimnion The lower, colder layers of water, beneath the **epilimnion**. There is not enough light for **photosynthesis** to occur and levels of dissolved oxygen are low.

hypothesis An idea or a suggestion or a theory which will be proved or disproved by studying the relevant facts. A problem-solving technique will be necessary to investigate the hypothesis.

I

ice Frozen water, or snow which has passed through the **firn** stage to form glacier ice. The relatively low density of ice means that it floats.

ice age A prolonged period of colder climatic conditions, during which snow and ice covered large areas of the Earth. There have been several ice ages in the past. The most recent began about 2 million years ago and is often referred to as the "Ice Age." During this period, ice advanced for several thousand years, and then retreated for about the same period. An advance of ice is called a "glacial phase," and the retreat is an "interglacial." We are living in an interglacial at present, although there are still ice age conditions over Greenland and Antarctica.

iceberg A floating mass of ice which originated on land as part of a glacier or ice sheet. As the ice moves slowly outward from a snowcap, some reaches the sea. It floats and is gradually broken off by the movement of tides and waves. Some icebergs may be several hundred feet in size, but they gradually melt and shrink as they float across the oceans. Icebergs from Greenland are quite tall, whereas those from the Antarctic are flat and tabular in shape. Up to 80% or 90% of an iceberg is below the surface of the water. As they melt, icebergs sometimes turn over; this readjusts the balance. One of the major areas for icebergs is on the route of the cold Labrador Current, which brings them southward along the coast of Greenland, toward Newfoundland.

icecap A large expanse of snow and ice covering a mountain range, an island or few square miles of land, as, for example, in parts of Iceland, some of the islands off the northern coast of Canada and Spitzbergen within the Arctic Circle. An icecap is a smaller version of an ice sheet.

icefall A section of a glacier where there is an abrupt change of gradient and the ice goes down a steep slope. As it does so, many transverse crevasses form, and at the bottom of the icefall there will be a mass of **séracs**, where the ice has been compressed again and the crevasses have closed up. An icefall is a very difficult area to cross and major icefalls in the Himalayas and Antarctica always create problems for explorers and climbers. Great Khumbu near Mt. Everest is a famous example, and there are many small examples in the Alps.

ice sheet A large expanse of snow and ice which covers a land mass; for example, Antarctica. During the Ice Age, an ice sheet covered much of Norway and Sweden, and spread across the North Sea to join the ice of Britain. It also covered parts of northern Germany and Denmark. At the same time, there was another ice sheet across Canada and the northern USA. Ice sheets can cover entire mountain ranges and can spread across lowlands and oceans. The ice may become very thick, and in Greenland there are places where the ice measures around 10,000 ft [3,000 m] in depth. Antarctica has areas of similar thickness. If all the ice melted from these two ice sheets, the level of the oceans could rise by as much as 300 ft [100 m].

ice shelf A sheet of floating ice on the sea. There are large areas of shelf ice around Antarctica, and small areas on the edge of the Arctic; for example, off the northern coast of Canada.

igneous A term denoting rocks of volcanic origin which were formed from **magma**. If the magma solidified beneath the Earth's surface, cooling would have been slow; if it solidified at the surface, cooling would have been more rapid. The faster the cooling process, the smaller the crystals which make up the rock. All igneous rocks are crystalline. In addition to grain size, the other important differences found in igneous rocks are related to the chemical content of the minerals in the

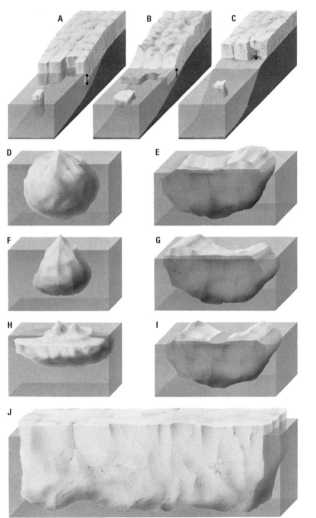

◀ **Icebergs** form in several ways. When a glacier reaches the sea it floats away from the bed. The movement of waves and tides exerts pressures on this floating ice causing lumps to break away (A). If the glacier is moving rapidly when it reaches the sea, a projecting shelf of ice forms under the water. The buoyancy of this shelf exerts an upward pressure causing pieces to break off (B). The snout of the glacier may be above the level of the sea and hence lumps may break off under the force of gravity and fall into the water (C). The forming of new icebergs is known as "calving". D–I show typical shapes of icebergs. Northern icebergs come from the Greenland ice sheet, but the largest ones originate in Antarctica (J). The largest iceberg ever seen was 336km (208mi) long and 97km (60mi) wide.

rock. Acidic minerals, including **quartz**, some **feldspars** and some micas, are generally light in color. Conversely, basic minerals are generally dark in color; they include some feldspars, biotite **mica**, hornblende and augite. Igneous rocks formed at depth are called plutonic, those formed at the surface are called volcanic and those formed between are hypabyssal. **Volcanic rocks** are extrusive, and the commonest variety is **basalt**. **Hypabyssal** rocks are found in **sills** and **dikes**, and they are intrusive rocks. **Plutonic rocks** form in large bodies of magma called **batholiths**, **stocks**, and "bosses" (circular intrusions, lying at a steep angle to the surface). The commonest plutonic rock is **granite**, which is acidic in mineral content; the basic equivalent is called **gabbro**. Igneous rocks do not have bedding planes, and they do not contain fossils. They are mostly quite resistant to erosion, and often form high ground.

illuviation The process in a soil whereby materials from upper horizons are leached out (i.e. washed down) and redeposited in lower horizons.

ILO SEE **International Labor Organization**

immediate runoff SEE **surface runoff**

immigrant A settler from another country or region. Immigration is generally used to describe international movements only.

impermeable A term denoting rocks which are nonporous and therefore do not absorb water. **Granite** is a good example of an impermeable rock; however, it is **pervious**, because water can pass through its

joints. An exception is clay, which is **porous**. It contains numerous tiny pores, but when these have been filled with water they prevent the movement of water through the rock, making it impermeable.

impervious A term denoting a rock which is not **pervious**. It will not allow the passage of water because it contains no joints, cracks or fissures; good examples are slate and shale.

import A commodity or article brought into a country from abroad. Imports may include minerals to be purified and processed, sources of fuel and energy, manufactured goods or foodstuffs. Imports generally have to be paid for out of the income from **exports**. Many **Third World** coun-

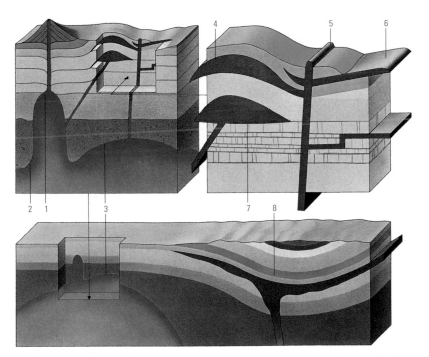

▲ **igneous rocks** Many different shapes of igneous rock can be found. Among the most common are the neck (1), which is a circular vertical feed channel of a volcano. A stock (2) is a large mass of rock that solidified at great depth. A batholith (3) is a large body of granite that has no detectable bottom. A laccolith (4, 7) is a dome-shaped mass that has forced the rock above it to arch. A dike (5) is a vertical, sheet-like mass of rock. A sill (6) is a horizontal, sheet-shaped body of rock. A lopolith (8) is a saucer-shaped mass.

tries export raw materials, but have to import most of their manufactured goods.

import substitution The replacement of goods that were previously imported by home-produced goods, thereby saving money and hopefully improving a country's balance of payments.

incised meander An old meander which has become deepened and has cut down into the landscape. This is generally the effect of extra erosion caused by **rejuvenation** of the river, sometimes as a result of **river capture** providing extra water, or of uplift of land near the source of the river producing an increased gradient. The extra erosion causes the river to flow at a level below the former **flood plain**. Incised meanders may be ingrown or entrenched. In an "ingrown meander" some lateral as well as vertical erosion has occurred, and the cross-section of the valley may be asymmetrical because of lateral erosion on the outside of the bend only. "Entrenched meanders" have been affected simply by vertical erosion, and the valley cross-section is even and symmetrical.

indigenous people People who are native to an area, i.e. they have not moved there from somewhere else. (The term "indigenous" can also be applied to plants, animals and rocks.)

industrial park A specially selected site, generally on the outskirts of a city, where various businesses and industries are located in new facilities. The park is generally purpose-built and will be well equipped with road transportation, electricity, water supply, etc. Industrial parks generally contain **light industries** only. The British name is "industrial estate."

industrial inertia SEE **geographical inertia**

industrial location Old **heavy industries,** such as steel manufacture, were always located on coal fields, especially if iron ore and limestone were available local-ly. Most new steel works are now located on the coast, so that materials can be transported by sea. The provision of transportation has become a more important factor than proximity to raw materials. In the case of some mineral purification, the industry is located near the mine; for example, copper and tin mining in Bolivia. In other cases, a mineral may be transported many miles for refining; oil from the Middle East is refined in Japan and Europe. Aluminum refineries are often located near sources of hydroelectricity, because the power is more significant than the raw material, as at Kitimat in British Columbia, western Canada, and Fort William in Scotland. Some industries are now located near their markets because they need only small quantities of raw materials, and labor and markets have become more important than raw materials as locating factors; for example, electrical and electronic industries near London, England. Sometimes locations may be chosen because of political decisions. Governments want to create jobs in areas of high unemployment, and so they often give tax incentives and grants to companies which will locate their factory in a place selected by the government. SEE ALSO **government incentives**

Industrial Revolution The period when the development of steam power and many other inventions enabled a great deal of rapid progress to be made in a variety of industries (also known simply as "industrialization"). This mainly took place in the years 1780 to 1820. As steam power developed and industries expanded, more workers were needed for more factories. This contributed to a rapid growth of wealth and of the urban population. The early British industries were located on coal fields; for example in South Wales, Yorkshire, Lancashire, northeast England and central Scotland. Steel, engineering, chemicals, cotton and woolens were the major developments. After the Industrial Revolution, there was continuous change in the industrial areas, as newer and more expansive developments took place.

infiltration The downward movement of water from the surface into the soil. The rate of infiltration depends on the amount of rainfall and its intensity, the vegetation cover, the compactness of the surface, the porosity and permeability of the rock, and the amount of water already in the soil.

informal economy Casual employment, often with irregular hours, using children and sometimes of dubious legality. This is found in many parts of the **Third World**, where there are many street traders. Small industries are increasing in many cities, as well as services such as shoe-cleaning. It is believed that the informal sector employs at least 50% of the workers in some large cities. There is also an informal economy in advanced countries as workers do part-time jobs while receiving state assistance, and some people work at two jobs, by doing evening or night shifts in addition to their normal working hours.

infrastructure A basic framework of roads, power and water supplies, schools, hospitals, etc.

ingrown meander SEE **incised meander**

inlier An area of old rocks surrounded by younger rocks. It sometimes consists of quite resistant rock which has been slow to erode. Sometimes an inlier is the result of erosion of an anticlinal structure; the exposed underlying rocks are surrounded by the younger rocks of the structure.

inner city A residential area in or near the center of a city. Many inner-city areas suffer from old and decaying housing, often in need of repair or renewal. Gradually, they are being redeveloped, but there are frequently cases of vandalism and other social problems, and considerable expenditure is needed to improve conditions in the inner cities, since land prices are very high. SEE ALSO **gentrification**

input Any of the items (energy, raw materials, etc.) which are put into a system. For example, in farming, seed, fertilizer and sunshine would be inputs, but in a steel works the major inputs would be iron, coal and limestone. All systems have inputs and outputs. SEE ALSO **output**

insolation The amount of energy received from the Sun. The amount varies according to the length of day, latitude and condition of the **atmosphere**. Polar regions receive long hours of insolation in summer, but none at all in the winter, while equatorial regions receive similar amounts throughout the year. Deserts have clear skies and receive more insolation than some equatorial zones, where cloud is present for a few hours each day. **Pollution**, as well as **cloud**, can restrict the amounts of insolation reaching the Earth.

instability The condition of an air mass with a lapse rate which is greater than the **dry adiabatic lapse rate** (about 1°F per 200 ft [1°C per 100 m]). If it rises, the air mass is likely to continue rising to such a height that condensation may occur. Therefore, instability of air means that rain is a possibility. When unstable air masses are passing overhead, cloud will form in many places, especially over mountains or areas where there has been extra heating. This means that instability is greatest during daylight hours, when more heating is likely to occur.

integration The process of carrying out all the distinct stages of an industry in one location. Large new steel works are generally integrated; raw materials go in one end of a large factory, and finished products come out at the other end. Many automobile factories are integrated. A conveyor belt moves from one end of the factory to the other, and completed cars can be driven away at the other end of the belt. Integration has also taken place in the cotton and woolen industries; the raw materials are cleaned, spun, woven and made into articles of clothing or other items, all within one factory.

intensive farming Any system of farming which gives a high return per acre.

Generally much labor, or capital, is necessary in order to produce high yields, and the land is cultivated annually and is not left fallow. Most parts of Britain and many parts of Europe are intensively farmed. Market gardening is one of the most intensive land uses, and greenhouse cultivation is especially intensive. Frequent, possibly daily, inputs are vital for the commercial and economic success of some intensive farms. A few areas of farm land are intensive yet subsistence, and this applies particularly to some of the rice-growing areas of Southeast Asia, such as the Si delta in southern China and the lowland areas in Java. SEE ALSO **horticulture**

interglacial SEE **ice age**

interlocking spurs A series of spurs on alternate sides of a river valley. They are the remnants of high ground which has largely been eroded by the river. The meandering course of a river causes erosion to take place on alternate sides of the valley. In some places where there has been no recent erosion, the higher ground will protrude as spurs. From the valley it is possible to see the spurs jutting out and overlapping, or "interlocking." This can be seen clearly on a small scale in the upper course of a river, but it also occurs in the middle course. By the time the lower course has been reached, the valley width will be too great for the spurs to overlap, and the valley walls are normally so reduced in height that any remaining spurs are quite low.

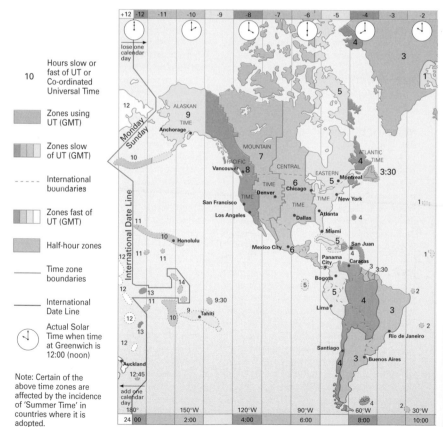

10 Hours slow or fast of UT or Co-ordinated Universal Time

Zones using UT (GMT)

Zones slow of UT (GMT)

International boundaries

Zones fast of UT (GMT)

Half-hour zones

Time zone boundaries

International Date Line

Actual Solar Time when time at Greenwich is 12:00 (noon)

Note: Certain of the above time zones are affected by the incidence of 'Summer Time' in countries where it is adopted.

intermediate technology SEE **appropriate technology**

intermittent stream A stream which flows for only part of the year. Intermittent streams dry up during the dry season, which is the winter in savanna regions and the summer in Mediterranean regions. Desert areas often have intermittent streams

▼ **international timezones** The Earth rotates through 360° in 24 hours, and so moves 15° every hour. The world is divided into 24 standard time zones, each centred on lines of longitude at 15° intervals. At the centre of the first zone is the prime meridian or Greenwich meridian. All places to the west of Greenwich are one hour behind for every 15° of longitude; places to the east are ahead by one hour

which flow for a few days or weeks after heavy rainstorms. SEE ALSO **exotic stream**, **perennial stream**

intermontane Between two mountains. The term is generally used to describe a **plateau** or a **basin**. An intermontane basin is a large hollow situated between two mountainous areas; for ex-

for every 15°. When it is 12 noon at the Greenwich meridian, 180° east it is midnight of the same day – while 180° west the day is just beginning. To overcome this, the International Date Line was established, approximately following the 180° meridian. Thus, if you travelled eastwards from Japan (140°E) to Samoa (170°W), you would pass from Sunday night into Sunday morning.

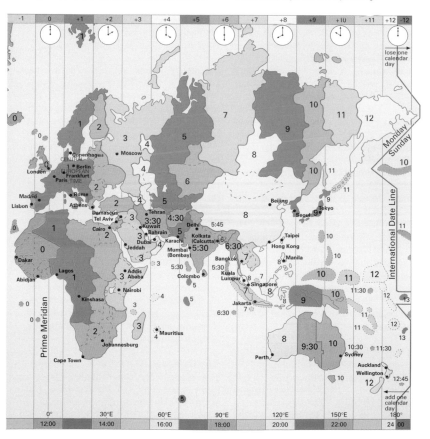

ample, Great Salt Lake in Utah, USA. An intermontane plateau is a high-level area which is lower than the surrounding land. The plateau of Tibet is much lower than the adjacent Himalayas, and in South America the plateaus of Peru, Ecuador and Bolivia reach over 10,000 ft [3,000 m], but are dwarfed by the Andean ranges alongside them.

International Date Line A theoretical line which approximates to the 180° line of longitude. It is used to overcome a time problem caused by travel; anyone crossing the line from west to east has to move the clock forward by 24 hours, and jump one day ahead, whereas anyone going from east to west puts the clock back 24 hours, and has to live the same day twice. The changes of time are caused by the Earth's rotation. As the Earth spins around, different parts turn to face the Sun and receive daylight. Traveling eastward from Greenwich, England, the time becomes later, at the rate of one hour per 15° of longitude. Moving westward from Greenwich, times are behind Greenwich Mean Time, by one hour per 15°, so that 15°W will be two hours behind 15°E, 30°W will be four hours behind 30°E, and so on. Eventually, there is a difference of 24 hours between 180°W and 180°E, but since the two lines of longitude are the same, the International Date Line was devised to solve the problem.

International Labor Organization (ILO) One of the specialized agencies of the United Nations. It is based in Switzerland and aims to improve international standards and social and economic conditions for employees worldwide.

Intertropical Convergence Zone SEE **doldrums**

intervisibility The visibility of one point on the landscape from another. It is possible to use contour maps to work out whether from one particular location it is possible to see other locations in the surrounding countryside. In order to determine whether hills, buildings or woods would block the view, it is necessary to draw a careful **cross-section**.

intrazonal soil A soil which develops in a particular environment, irrespective of climatic conditions. Examples include **rendzinas**, which develop on limestones, and **gley** soils, which develop on water-logged land.

intrusion The process by which "intrusive rocks" are formed.

intrusive rocks SEE **extrusive rocks**

Inuit The Eskimo people who live in Greenland, northern Alaska and northern Canada. Traditionally they lived in small, isolated communities as nomadic hunters, but gradually started moving to towns. In 1999, their own territory, Nunavut, was created. SEE ALSO **Nunavut**

inversion layer SEE **temperature inversion**

inverted relief A type of relief in which former uplands have been so eroded that they now lie below the surrounding areas which were previously lower. When folding takes place, the upfolded rocks in the **anticline** are often stretched and cracked, which means that they are more easily eroded than the more compressed and consolidated rocks in the **syncline**. Erosion over a few thousand years can wear the anticlinal area down to a lower level than the syncline.

invisible exports Items in a country's international trade which earn foreign exchange without the transfer of goods out of the country to another. For example, services such as banking and insurance, or shipping, air freight and expenditure by tourists visiting the country.

ionosphere The zone in the **atmosphere** above the **thermosphere** with a range of about 60 miles to 370 miles [100 km to 600 km] in altitude. Here gas molecules, mainly helium, oxygen and nitrogen, are electrically charged by the sun's

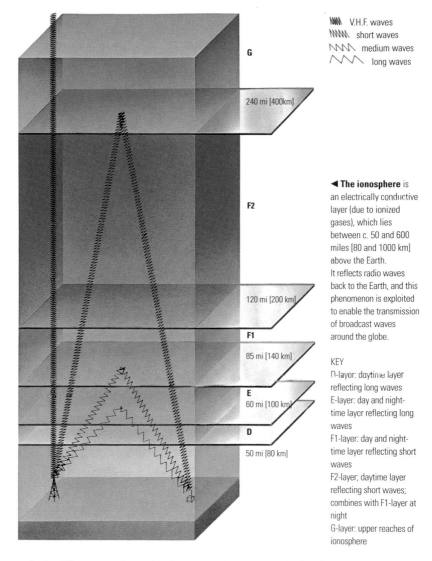

WW V.H.F. waves
WWW short waves
WWN medium waves
WW long waves

◄ **The ionosphere** is an electrically conductive layer (due to ionized gases), which lies between c. 50 and 600 miles [80 and 1000 km] above the Earth.
It reflects radio waves back to the Earth, and this phenomenon is exploited to enable the transmission of broadcast waves around the globe.

KEY
D-layer: daytime layer reflecting long waves
E-layer: day and night-time layer reflecting long waves
F1-layer: day and night-time layer reflecting short waves
F2-layer; daytime layer reflecting short waves; combines with F1-layer at night
G-layer: upper reaches of ionosphere

radiation. They group themselves into layers which reflect radio waves of differing frequencies. The high energy of ionospheric gas gives it a notional temperature of more than 3,500°F [2,000°C], although its density is negligible.

Iron Age The period in human development following the **Bronze Age**, when iron was smelted and used for tools, utensils and weapons. The Iron Age reached Britain from Europe in about 500 BC.

Iron Curtain A term used first by Winston Churchill in 1946 to describe the imaginary barrier between Western Europe and the USSR and Eastern bloc countries (Poland, East Germany, Czechoslovakia, Hungary,

Romania, Bulgaria and Albania). It ceased to exist as an accurate description in about 1989, with the reunification of Germany and political changes in Eastern Europe.

Iron and steel A **heavy industry** which for a long time has been regarded as a vital part of any nation's industrial development. During the past 20 to 30 years, Japan, Taiwan, South Korea and many other countries have developed steel industries; as a result, the American and European steel industries have declined because of the increased competition.

iron pan SEE **hardpan**

irrigation The supplying of water by artificial means to an area where there is a shortage, usually for the purpose of growing crops. Various methods are employed. Simple, old-fashioned methods like the **Archimedes' screw** and the **shaduf** are still useful for lifting water from one level to another, and water wheels are used for the same purpose. Wells are increasingly numerous in many parts of the world, and simple wells, which are merely holes in the ground, are being replaced by tube wells which are lined and can reach deeper supplies of water. There is often a surfaced area around the top of the well in order to reduce erosion by humans and animals. Most important nowadays are the large

schemes for irrigation in which water is available throughout the year. This is called **perennial irrigation**, and requires a dam and a large lake in which vast quantities of water can be stored. Some of the largest lakes, such as Lake Nasser behind the Aswan dam in Egypt, contain enough water to last for two years, even if there was no more rainfall. The dams often supply hydroelectric power as well. Leading from the lake will be a network of canals, which not only provide reliable supplies of water throughout the year, but also enable a greater area of land to be irrigated. Some of the major irrigated areas are in deserts; for example, the Nile valley in Egypt and the Indus valley in Pakistan. There is a large amount of irrigation in regions with a **Mediterranean climate**, including southern France and California, USA. Even quite wet areas, such as parts of India and China, can benefit from irrigation in order to guarantee reliable supplies of water, or to enable a second crop to be grown in the drier winter months, when temperatures are still high enough for rice and other plants to grow successfully.

island An area of land completely surrounded by water.

isobar A line on a weather map joining places of equal pressure. The pressure readings are normally taken at or near sea level, although upper atmosphere pressure readings are used to help with weather forecasting and also in connection with planning routes for aircraft. Isobars drawn on **synoptic charts** indicate weather patterns, including the location of **depressions** and **anticyclones**. Average isobars can be drawn on continental or world maps to show the generalized pressure distribution. On daily weather maps, the isobars are usually drawn at intervals of 4 mb.

isohyet A line on a weather map joining places with equal rainfall. Isohyets are generally drawn to show the distribution of rainfall over a year, or perhaps for the summer months, but they may be drawn for any period of time.

WORLD TOP TEN ISLANDS (by order of size)			
Name	Location	Size (km^2)	Size (mile^2)
Greenland	Greenland	2,175,600	839,800
New Guinea	Indonesia	821,030	317,000
Borneo	S.E. Asia	744,360	287,400
Madagascar	Indian Ocean	587,040	226,660
Baffin Is.	Canada	508,000	196,100
Sumatra	Indonesia	473,600	182,860
Honshu	Japan	230,500	88,980
Great Britain	UK	229,880	88,700
Victoria Is.	Canada	212,200	81,900
Ellesmere Is.	Canada	212,000	81,800

isoline A line on a map which is joining places having the same value of any selected element. An **isobar** or **isohyet** is a type of isoline, and so too is a **contour**. ALSO CALLED **isopleth**

isopleth SEE **isoline**

isotherm A line on a weather map joining places of equal temperature. Before maps are drawn, temperatures are normally reduced to their sea-level equivalent. They can be drawn to show the average January or July temperatures, or to show the temperature at a given time on any particular day. Thus, they can show precise temperature information, or give averages over a longer period of time.

isostasy The state of balance or equilibrium of the Earth's land masses. The continental rocks, which are made up of the **sial**, sit on top of the denser rocks, the **sima**, which make up the ocean floors. When the continents are large they weigh down into the sima, but as erosion reduces their weight, they gradually rise. This is "isostatic readjustment." During ice ages, the weight of the ice presses the continents down. As soon as some ice melts, the land masses begin to rise because of isostasy. When this happens, the former coastlines are elevated to form **raised beaches**, such as those which can be seen along the Scandinavian coastline.

isotropic Showing the same physical proportions in every direction. The idea of an "isotropic plain" was used by many people, such as **Weber** and **von Thunen**, when developing their land-use models. It is rare, however, to find any area which is truly isotropic, although flat, reclaimed land, such as in the polders in the Netherlands or the Fens in England, is nearly isotropic.

isthmus A narrow strip of land, with water on either side, which connects two large areas of land; for example, the isthmus of Panama connects the Americas.

J

jarrah One member of the *Eucalyptus* tree genus, found in Western Australia.

jet stream A strong current of air blowing through the **atmosphere** at a high altitude. The main jet streams are in the middle and subtropical latitudes. The temperate jet stream flows from west to east, though along a wavelike route, going north and south as it travels all around the world. It can blow at speeds of up to 230 mph [370 km/h] at times, and so is an important factor for aircraft. If planes can fly with the jet stream, which often blows at a height of 30,000 ft to 40,000 ft [10,000 m to 13,000 m] above the Earth's surface, it can increase the speed of the journey and save fuel costs. If the aircraft is going westward across the Atlantic Ocean, it will try to fly above or below the jet stream. Jet streams influence the weather systems in the lower layers of the atmosphere, as **depressions** tend to follow their route. Jet streams change position fairly slowly, and once established in a particular location may persist for a week or two.

▼ **Jurassic** By the Jurassic period, the Gondwanaland landmass had begun to break into separate continents.

joint A vertical or near-vertical crack in a rock. Joints occur in **sedimentary rocks** because of shrinkage or weathering along a line of weakness. In **igneous** rocks they are formed as a result of cooling and contraction, as **magma** solidifies. Joints are much smaller than faults and there is little or no movement on opposite sides of the joint; they are often a site of **weathering** and **erosion**. Cliffs and steep slopes may be influenced by the positioning of joints. Joints are very frequent in **Carboniferous** limestone and also in **granite** and **basalt**. Basalt often cools into hexagonal-shaped blocks because of the vertical cracks which occur during cooling. A fine example of such columnar jointing in basalt is the Giant's Causeway in County Antrim, Northern Ireland.

joule A very small unit of energy. One joule equals 0.239 calories; one megajoule (MJ) equals 239 calories; and one gigajoule (GJ) equals 239,000 calories. About 10.5 MJ, or 2500 calories, are needed as the daily food requirements of an average working person. The amount of solar radiation can also be measured in joules. Aberystwyth in central Wales receives 210 joules per sq cm [1,350 joules per sq in]

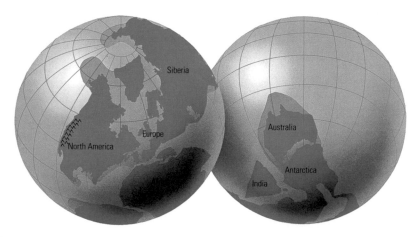

Siberia

Europe

North America

Australia

Antarctica

India

per day in the winter, but 1,880 joules per sq cm [12,130 joules per sq in] per day in summer. Brisbane in Australia varies from 1,150 to 2,480 joules per sq cm [7,420 to 16,000 joules per sq in] per day and Algiers from 880 to 2,800 joules per sq cm [5,680 to 18,000 joules per sq in] per day.

junction In **geomorphology**, the meeting point of two rivers.

jungle A type of tropical or monsoon forest with dense undergrowth. Many tropical forests, which have rain all through the year, have so much growth of vegetation that there is a dense canopy through which sunlight cannot penetrate. In these areas near the Equator there is no undergrowth, and so no jungle. It is where there is a short dry season and the forest is less dense and more open that undergrowth flourishes. The most impenetrable areas for jungle growth are in the monsoon regions of Southeast Asia, though even here large areas of woodland are not jungle. SEE ALSO **tropical rain forest**

Jurassic A geological time period which occurred about 150 million years ago. North America separated from Africa during this period and the Rockies were created by folding. Jurassic rocks are clays, sandstones and limestones. Many fossils are found in Jurassic rocks, including ammonites, belemnites and some dinosaurs. SEE ALSO **geological column, geological time scale**

juvenile water Water that is found within **magma** and appears at the Earth's surface for the first time after a volcanic eruption.

K

Kainozoic SEE **Cenozoic**

kame A **fluvioglacial** deposit formed by streams flowing on to or beneath the ice. Streams flowing down mountain sides on to glaciers deposit sand and gravel at the point where the stream first reaches the ice, or where the water goes down into a **crevasse**. The accumulations of sand and gravel gradually fall down to the valley floor as the ice melts. Some kames are the result of deltaic formations created by subglacial streams. Kames are generally small features, 60 ft to 100 ft [20 m to 30 m] in length or breadth, though occasionally "kame terraces" form along the side of the ice and they may be elongated.

kanat SEE **qanat**

▼ **kame** Glacial meltwater sometimes deposits terraces along the side of a glacier. Such terraces are deposited most commonly during the disintegration of a glacier. Thus most kame terraces were formed at the end of the last ice age.

kaolin SEE **china clay**

karez SEE **qanat**

karst scenery The type of landscape found on limestone where most of the water has gone underground. "Karst" takes its name from the Krast (German: *Karst*) region of Croatia and Slovenia. **Solution** weathering wears away much of the limestone, and rivers also erode the rock. Most rivers are underground, and there are caves and large caverns. The largest caverns may collapse to form **gorges**, and gradually the entire area of limestone may be worn away. The surface features of the limestone usually include **clints, grikes, limestone pavements** and **swallow holes**.

katabatic A term denoting a wind which blows downhill. Katabatic winds are most likely to occur at night when air which has been cooled by radiation accumulates in the valleys and begins to flow downhill. They can also be found near **glaciers** and **ice-caps**

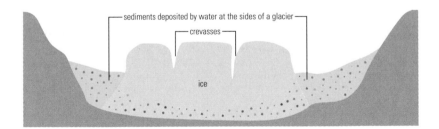

sediments deposited by water at the sides of a glacier

crevasses

ice

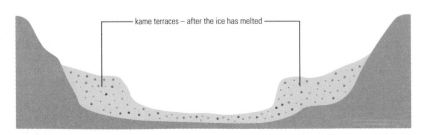

kame terraces – after the ice has melted

KARST SCENERY

Karst landscapes, such as those of the Appalachian Mountains, Dinaric Alps, and areas of southern China occur because beds of limestone rock, composed primarily of the mineral calcite (calcium carbonate), are progressively eroded by water. Carbon dioxide from the atmosphere dissolves in rain to form weak carbonic acid, which dissolves the rock, particularly along joints and bedding planes, enlarging them to produce sink holes (A). Dolines (B) are produced by cave roofs falling in; they can be enlarged to produce gorges (C), finally leaving the characteristic uneroded limestone cores of the Karst landscape (D).

when the cold air from above the ice starts to move downhill to lower levels. Winds of this type may blow during the day as well as at night; for example, in Greenland. The opposite of katabatic is **anabatic**. Winds that are anabatic blow uphill because of heating during the day. They are less powerful and less frequent than katabatic winds.

kettle hole A small hollow 30 ft to 60 ft [10 m to 20 m] across, situated in morainic debris or on an **outwash plain**. A kettle hole is formed when a block of ice that has been covered with glacial deposits slowly melts; as it does so, the remaining debris collapses to form a hollow. Kettle holes are often marshy and may even contain a small pond.

khamsin A hot, southerly wind which blows across Egypt toward the Mediterranean Sea. It is very dry, with relative humidity as low as 10% on occasions. This will crack the skin, as well as woodwork. It blows into the front end of a depression that is passing along the Mediterranean. Large quantities of dust are sometimes carried over long distances by the khamsin, which is a similar wind to the **sirocco** of Algeria.

kibbutz A type of cooperative farm and settlement found in Israel. All the members of the kibbutz, help with all kinds of work, whether farming, cooking or cleaning. Many kibbutzim have been created in the southern regions of Israel on land reclaimed from the desert by means of extensive irrigation.

kidney ore SEE **hematite**

knick point A break in the **long profile** of a river which may be as a result of **rejuvenation** caused by uplift of the land.

Alternatively, it can result from **river capture**. It is normally marked by a small fall or some turbulent water, but in time the knick point will be smoothed off by river erosion. ALSO CALLED **nick point**

knoll A small rounded hill.

knot A measure of speed, sometimes used instead of miles per hour. It is slightly faster than miles per hour, as it represents one nautical mile per hour. It is a relic from the days of sailing ships.

kolkhoz (*Russian*) A collective farm.

L

labor In economics, a factor of production involving manual or intellectual work. The "quantity" of labor is the amount of work done in terms of production or time, while the "quality" of labor shows the degree of skill and intelligence required (i.e. it may be skilled, semiskilled or unskilled).

labor intensive A term used to denote an industry or business activity which uses proportionately more labor than capital or machinery. This can apply to agriculture as well as industry. In some areas, most of the farm work is done by hand, as, for example, in many of the rice-growing areas of Asia. Some factories still employ a large labor force; for example, some confectionary manufacturers or toy makers, but many large factories are now capital intensive and use machinery and automation. Labor-intensive methods are probably more appropriate in the less developed countries, as they have large supplies of labor available but not much money to buy machinery. SEE ALSO **capital intensive**

Labrador Current A cold sea current found in the North Atlantic Ocean.

laccolith A **igneous** intrusion which has forced its way between the existing strata and is therefore concordant with the structure. Often the overlying rocks bulge upward, forming a small hill on the surface. Laccoliths are formed of **magma**; they are smaller than **batholiths**, and generally have a flattened base and a domed upper margin.

lacustrine Relating to a lake. Lacustrine deposits are those which have been laid down on the bed of a lake. When the lake is filled up and disappears, which happens to all lakes eventually, a flat and fertile lowland will remain. The Red River valley south of Winnipeg in Manitoba, Canada, consists of lacustrine deposits from a post-glacial lake, Lake Agassiz. In Britain, the Vale of Pickering in Yorkshire was formerly a lake, and many of the valleys in the Lake District, such as Langdale, formerly contained lakes.

lagoon A stretch of shallow water partially or completely cut off from the sea. If it is completely cut off, there will be some percolation of water through the ridge or **spit** which separates the lagoon from the sea. Lagoons are often found behind spits, bars and **coral reefs**.

lake A broad term that describes a body of water (usually of some size) lying in a depression in the Earth's surface. A particularly large lake, that is natural and saline, may be called a "sea" (for example, the Dead Sea); while a tiny, and also natural, lake is known as a "pond" or "pool." Lakes may have an in- or outflowing river and are not always permanent features on the landscape.

land breeze A light wind blowing seaward from the adjoining land, usually at night. During anticyclonic conditions, when local weather controls develop, the land cools down more quickly at night than the

WORLD TOP TEN LAKES (by order of size)			
Name	Location	Size (km²)	Size (miles²)
Caspian Sea	Asia	371,800	143,550
Lake Superior	Canada/USA	82,350	31,800
Lake Victoria	E. Africa	68,000	26,000
Lake Huron	Canada/USA	59,600	23,010
Lake Michigan	USA	58,000	22,400
Aral Sea	Kazakhstan/ Uzbekistan	36,000	13,900
Lake Tanganyika	C. Africa	33,000	13,000
Great Bear Lake	Canada	31,800	12,280
Lake Baykal	Russia	30,500	11,780
Lake Malawi/ Nyasa	E. Africa	29,600	11,430

sea. Air starts to rise from the relatively low pressure of the warm sea, and cool air drifts out from land to sea, until the sun has warmed up the land in the morning. During the day a reverse flow, or sea breeze, will develop. Land and sea breezes both develop because the land heats up and cools down more rapidly than the sea. In some tropical areas, where land and sea breezes are frequent and powerful, fishermen can go out to sea with the land breeze very early in the morning; by midafter- noon, a sea breeze will have developed to blow them back to shore. SEE ALSO **sea breeze**

landform A landscape feature, such as a hill, plateau or valley. The study of landforms is called **geomorphology**.

landlocked An area, often a country, that has no direct access to the sea. There are 38 landlocked countries in the world, from Andorra to Zimbabwe. Landlocked coun-

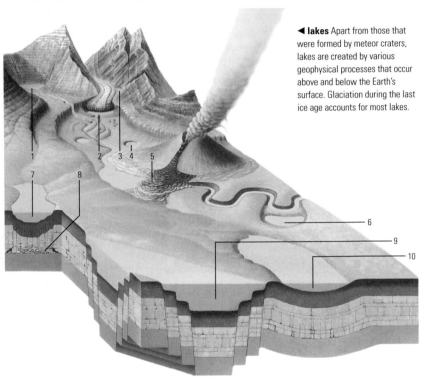

◀ **lakes** Apart from those that were formed by meteor craters, lakes are created by various geophysical processes that occur above and below the Earth's surface. Glaciation during the last ice age accounts for most lakes.

A cirque lake (1) forms as water collects in the hollow left at a glacier head. At the snout of a glacier, a lake may form as meltwater is trapped by the terminal moraine (2), a bank of rock and debris left behind as the glacier retreats. Lakes may also form behind lateral moraines (3) at the glacier's sides. Blocks of ice buried in glacial debris eventually melt, leaving hollows that become kettle lakes (4). Lakes may also be created by volcanic activity, as when larva flows into a river and solidifies into a dam (5). Subsequent cooling of the volcanic cone may permit a crater lake to form in the caldera. An oxbow lake (6) forms when accumulated sediment cuts off an exaggerated loop of a river. In regions where the underlying rock is limestone, a surface lake (7) may form in the sinkhole created by the collapse of an underground cave system (8). Large-scale earth movements may form a lake by creating surface depressions in which water can accumulate. Faulting of the rock strata (9) tends to produce deeper lakes than folded rock strata (10).

LAND AND SEA BREEZES

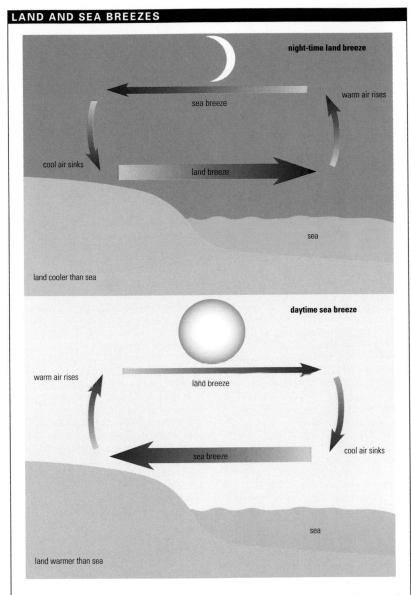

night-time land breeze

warm air rises

sea breeze

cool air sinks

land breeze

sea

land cooler than sea

daytime sea breeze

warm air rises

land breeze

cool air sinks

sea breeze

sea

land warmer than sea

▲ **land and sea breezes** Land and sea breezes occur because land warms up and cools down more quickly than water. At night, the land cools causing cool air to sink, creating higher pressure at low level. At the same time, relatively warm air over the sea rises, creating lower pressure. Air at ground and sea level thus flows towards the sea. At higher levels, air flows in the opposite direction. During the day, the process is reversed as the land warms more rapidly than the sea in the heat of the Sun.

LANDSLIDE

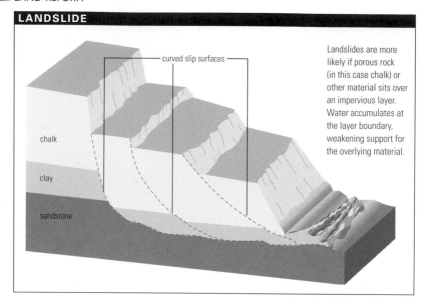

curved slip surfaces

chalk

clay

sandstone

Landslides are more likely if porous rock (in this case chalk) or other material sits over an impervious layer. Water accumulates at the layer boundary, weakening support for the overlying material.

tries often try to gain access to the sea; for example, along a river or a corridor of land.

land reform The redistribution of land. A common land reform is the breaking up of large estates and redistribution of the land among the peasants who were previously landless laborers. Land reform endeavours to reduce inequalities, since in many countries most of the land is owned by only a few people. There have been revolutions in several countries because of this unequal distribution. In many cases, the land reform has helped small farmers to grow more food and also earn money with cash crops. In other places the new landowners have not really known what to do with the extra land. Help and advice is often an essential part of any land reform programme. A totally different type of land reform has been taking place in parts of Africa, Cyprus, Germany and other countries. Here there are many farms which are fragmented, each farm consisting of several tiny plots of land. Attempts have been made to consolidate the farms and arrange for all parts of the farm to be together, rather than dotted about over several miles of countryside.

landscape The scenery, including the geomorphological features as well as the man-made features, such as fields, buildings, roads and cities.

landslide A usually rapid movement of rocks, soil and vegetation down a slope. It may be caused by an **earthquake**, but is generally the result of rain soaking the ground; the effects are exaggerated if there is an **impervious** layer beneath the surface. In colder climatic areas, the saturated rock may be affected by freezing and thawing of the water, and this increases the likelihood of landslides. ALSO CALLED **landslip**

landslip SEE **landslide**

land tenure The system of land ownership or rental. This has included sharecroppers in the USA and many parts of the world. SEE ALSO **sharecropping**

land use The ways in which land is used can be plotted on maps. Land utilization surveys have been conducted in Canada and the USA, as well as in other parts of the world. Land-use maps can also be drawn for **urban** areas; they can reveal zones of industrial use, business use, expensive housing, etc., and on large-scale maps it is possible to show the use of each building.

land-value gradient The scale on which the average value of land declines in proportion to its distance from a city's downtown area. There are exceptions to this generalization, and a line graph can be drawn to show the variations in land value. There is a general decline, but small peaks will appear at important crossroad areas and near suburban shopping centers.

lapilli SEE **pyroclastic rock**

lapse rate The rate at which temperature decreases with height above the ground. The average lapse rate is about 0.3°F per 100 ft [0.6°C per 100 m], and this is called the "environmental lapse rate." The rate will vary according to the water vapor content of the atmosphere. If dry air is rising, it will cool more quickly, at the **dry adiabatic lapse rate** (about 1°F per 200 ft [1°C per 100 m]). If the air is wet, the cooling rate will be slower, at the **saturated adiabatic lapse rate** (on average, about 0.25°F per 100 ft [0.5°C per 100 m]). The reason for the lower rate of decrease is that the water vapor gives off latent heat as it condenses, and this extra heat slows down the temperature change.

lateral erosion Sideways erosion. This occurs in rivers on the outside of **meanders**, and gradually the valley becomes wider. Lateral erosion eats into bluffs and slowly wears them away. **Corrasion** is the main process involved, but there may also be some **hydraulic action** and **corrosion**.

lateral fault SEE **tear fault**

lateral moraine A linear pile of rocks deposited by a **glacier**. As the glacier moved, rock fragments were pushed out to the sides, where they were left when the ice melted. In a **U-shaped valley,** the moraine may also include rocks which slipped down the valley walls – when the ice melted, the moraine slumped downward to the valley

floor. Usually, it will have been added to by the postglacial **scree**, which will also have accumulated at the foot of the valley wall.

laterite A reddish hard-baked soil which is found in tropical parts of the world. It results when hot wet conditions wash away most of the nutrient content of the soil, leaving behind hydrated oxides of iron and aluminum. The top is baked hard by the sun, and this factor, together with the leached nature of the soil, means that it is poor for agriculture. However, some laterites contain sufficient iron to be of commercial value. Laterites usually form when the soil has become exposed by forest clearance or by shifting cultivators who have not allowed the forest adequate time to recover.

latex A milky white fluid which is the source of rubber. It is obtained from various trees, especially *Hevea brasiliensis,* by tapping; that is, by making small cuts in the bark of the tree, through which the latex oozes. It is collected in a small cup which is attached to the tree just below the cut. SEE ALSO **rubber**

latifundium (*plural* latifundia) A large farm or estate in Spain and South America that is farmed extensively. Labor is usually employed on a short-term basis, often seasonally, or there may be a system of sharecropping. For a comparison, see **minifundio** and **huerta**.

latitude The distance north or south of the Equator, measured at an angle from the Earth's center. All lines of latitude are parallel to the Equator which is the zero line of latitude. Each degree of latitude is approximately 69 miles [111 km]. The tropics of Cancer and Capricorn are 23.5° away from the Equator, and the Arctic and Antarctic circles are at 66.5°, which is 23.5° away from the poles.

Laurasia The old continent which consisted of North America, Europe and Asia. Continental drift and plate tectonics caused the continent to split up into the separate continents which exist today. As the split occurred there was volcanic activity, for example in northern Ireland and western Scotland, similar to the present activity along the Mid-Atlantic (or Laurentia) Ridge in Iceland and elsewhere. SEE ALSO **continental drift, plate tectonics**

lava Molten rock or **magma** which has reached the Earth's surface as a result of volcanic activity. Lava varies in chemical content. If it is acidic it will be viscous and slow-flowing, but if it is basic it will be more fluid. Basic lavas form lava flows and often erupt out of fissures, as well as volcanic craters. Basic lavas are commonest at points where plate margins are moving apart, but acidic lavas occur at colliding plate margins and are often associated with very explosive volcanoes. Once they reach the surface, lavas cool and solidify. If the lava is very gaseous, a rough, jagged surface will result, which is described as "scoriaceous." SEE ALSO **aa, pahoehoe**

lava flow A movement of lava from a volcano or fissure. The movement may be fast if the lava is basic, but slow if it is acidic. A small lava flow can come from a volcanic crater, but sheets of lava may emerge from fissures. SEE ALSO **acid lava, basic lava, viscous lava**

lava plateau A large elevated area which has been built up by a series of volcanic eruptions, generally from fissures. A lava plateau covers several thousand square miles in the Deccan regions of India, and another large lava plateau is the Snake/Colombia area of the northwest USA.

lea Arable land temporarily sown with grass. Leas provide rich pasture for sheep and cattle.

leaching The process by which chemicals and nutrients are removed from a soil. Rainwater, especially in warm climatic regions, will dissolve anything soluble and wash it down from the **A horizon**. Once removed, these solubles can only be replaced very slowly. As a result, leached soils become coarse and are infertile. Large

areas of the tropical lands have become seriously leached because of the removal of the forest cover.

least cost location An industrial site selected for ease of access to the necessary raw materials. This will save transportation costs, and has been especially important in the iron and steel industry. Steel used to be made on coal fields to avoid transporting coal very far. Nowadays, because of changing technology, iron is the bulkier commodity. Many steel works in foreign countries are situated along the coast, such as those in Baltimore, Maryland, USA, and Osaka, Japan.

lee The sheltered side, i.e. from the wind.

legume Any plant of the family *Leguminosae,* such as peas, beans, clover and alfalfa, which extract nitrogen from the air and transfer it to their roots. When the plants die, the nitrogen is released into the soil. Legumes are sometimes described as nitrogenous; they are very beneficial to the soil, and are excellent crops in a rotation system.

leisure The use of free time is an increasingly important industry, and leisure has been a growth area of employment. Many leisure activities are located in cities; for example, sports centers, playing fields, entertainments, restaurants, etc. Others are located in rural areas, where they sometimes come into conflict with farming. Walking may create problems of erosion on footpaths, as well as damage to fences and walls. Many areas are now being set aside specially for leisure activities; for example, golf courses, country parks, theme parks, water sports, etc. Other important areas include **national parks** and **heritage coasts**, which provide space and opportunities for leisure, while remaining agricultural land.

less developed countries The poorer countries of the world, where agriculture remains the major activity and employer of labor. The beginnings of industrial development can be seen in most countries, but it is often only in the capital city and a small number of other cities. The less developed countries are sometimes referred to collectively as **The South**, and include much of Latin America, Africa and Asia.

leveche A hot, dry southerly wind which blows into southern Spain. It is similar to the **sirocco** and **khamsin**, and blows in advance of a **depression** passing along the Mediterranean.

levee A natural embankment formed alongside a river by the deposition of silt when the river is in flood. Silt is deposited all over the **flood plain**, but most is deposited near to the river banks, and so a slightly higher area is created alongside the river. Levees can help to prevent flooding, and are sometimes built up and strengthened artificially. Some levees grow so large that floodwater cannot return to the main channel. When this happens the river may change course, as has often occurred on the River Huang He (Hwang Ho) in northern China. On large rivers, such as the Mississippi in the USA, levees may reach heights of 50 ft [15 m].

ley SEE **lea**

liana A climbing plant which grows on trees in tropical forests. The stem of the liana may grow up to the top of the host tree, eventually killing it off. Lianas may be used as ropes by people living in the forests. ALSO CALLED **liane**

liane SEE **liana**

life expectancy The number of years a person may be expected to live based on statistical probability. Improvements in diet, hygiene and medicine have helped to increase life expectancy in most countries, although the more advanced countries are still much more favored than the less developed countries. In the countries of North America and Europe, life expectancy is commonly about 74 years for men and 79 years for women, whereas in several African countries the life expectancy is only about 48 years, although it

is always slightly higher for women than for men. The highest mortality rate is before the age of one. This means that the life expectancy of anyone over one year of age will be higher than the national average. Life expectancy becomes greater with age.

light industry The type of industry which produces light, often small goods, that can be transported easily by road. Light industries are often located on industrial parks, or alongside main roads, such as bypasses and beltways. Light industry is generally quite clean and nonpollutant, and can employ both male and female labor. Light industries often produce consumer goods, such as electrical goods, clothing and foodstuffs.

lightning A flash of light caused by a strong discharge of electricity in the atmosphere. The electricity is formed by tiny charges given off by large raindrops in a tall **cumulonimbus** cloud. Lightning occurs in thunderstorms, and it is the flash of lightning passing through the air which causes the **thunder**. The speed of light is much faster than the speed of sound, and so the lightning appears to precede the thunder. Lightning is visible on its way up from Earth after it has struck. So, if you see lightning, it has already struck, and missed you. Cloud-to-ground lightning looks forked, but cloud-to-cloud lightning often creates a sheet of light ("sheet lightning"). SEE ALSO **thunderstorm**

lignite SEE **brown coal**

limestone A **sedimentary rock** formed on the bed of a warm sea by an accumulation of dead sea creatures. At least 50% of a limestone consists of calcium carbonate. There are many different types, including **Carboniferous** limestone, chalk, dolomitic limestone, oolitic limestone, shelly limestone and Wenlock limestone. They all contain fossils or fragments of fossils. Because of the calcium carbonate content, limestones are soluble and **permeable**, and generally give rise to dry landscapes. **Chalk**

LIGHTNING

A Once an electrical potential of about 300,000 volts/ft (a million volts/m) has been created in a thunder cloud, the lightning process begins. A stream of electrons flows down, colliding with air molecules and freeing more electrons, and in the process giving the air molecules a positive charge (ionizing the air). This intermittent low-current discharge forms a highly branched pathway and is called the stepped leader (1). As the leading branches of the stepped leader, carrying large negative charges, near the ground they induce short upward streamers of positive electrical charges from good conducting points on the ground (2). When a branch of the stepped leader contacts an upward positive streamer, a complete channel of ionized air has been created. This allows a huge positive current called the return stroke to flow upward in to the cloud in the form of a bright lightning stroke (3). The different strokes have been given different colours to distinguish between them.

In nature, all lightning is colourless. The return stroke causes the first of the shock waves we hear as thunder (B). The flash effectively reaches the eye instantaneously, whereas the sound of thunder travels at approximately 1,100ft/s (330m/s). Therefore, the number of seconds between the flash and the thunder multiplied by 1,100 tells us how far away the lightning stroke is in feet. A fraction of a millisecond after the return stroke (3), a negatively charged dart leader passes down the ionized channel (4) and triggers another upward return stroke.

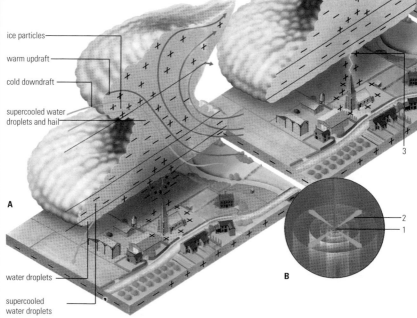

ice particles

warm updraft

cold downdraft

supercooled water droplets and hail

A

water droplets

supercooled water droplets

B

is pure limestone and forms rounded hills and vertical cliffs, if exposed along a coast. Chalk hills are often called **downs** or **wolds**. Carboniferous limestone is hard and forms high ground. Oolitic hills are generally more gentle than those of Carboniferous limestone, but have a very pronounced west-facing scarp. Carboniferous limestone, like chalk and oolitic limestone, has numerous **dry valleys**, but also contains caves with **stalactites** and **stalagmites**. SEE ALSO **oolite**

limestone pavement An outcrop of **Carboniferous** limestone in which horizontal rocks have been weathered into **clints** and **grikes**. These look vaguely similar to the slabs of a sidewalk, though the cracks between the rocks are wider and deeper. Clints are rectangular in shape and may be up to 6 ft [2 m] long, while grikes may be over 3 ft [1 m] in depth.

line graph A simple graph in which the values recorded are joined together by a continuous line, as, for example, a temperature graph. It is a very clear method of displaying information.

linear settlement An elongated settlement. It is usually **urban**, stretched alongside a route, such as a road or river, to take advantage of the transportation possibilities.

literacy rate The percentage of people in a given population who have been taught how to read and write. The literacy rate ranges from nearly 100% in most Western countries to less than 50% in several African countries.

lithosphere The solid layer which forms the surface of the Earth. It rests on top of the **mantle**, and contains the **sima** and the **sial**. Between the lithosphere and the mantle is the "Moho." ALSO CALLED **crust**. SEE ALSO **Mohorovičić discontinuity**

llanos (*Spanish*) Grasslands, especially the savanna-type grasslands of the Orinoco basin in Venezuela.

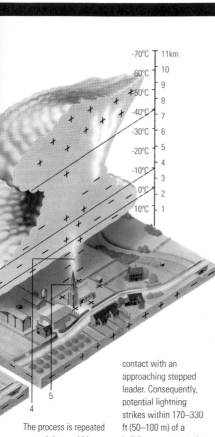

The process is repeated several times within fractions of a second until the charge in the cloud is neutralized. Lightning conductors (5) generate a strong positive streamer which encourages electrical contact with an approaching stepped leader. Consequently, potential lightning strikes within 170–330 ft (50–100 m) of a building are attracted to a lightning conductor. Return strokes and dart leaders are safely routed to the Earth along a wide copper strip with one end buried in the ground.

B Thunder is caused by the narrow lightning stroke heating the column of air (1) surrounding it to around 55,000°F (30,000°C), expanding it explosively (2) at supersonic speeds under a force of 10–100 times normal atmospheric pressure. The immense shock wave becomes a sound wave within about a metre, producing the sound of thunder.

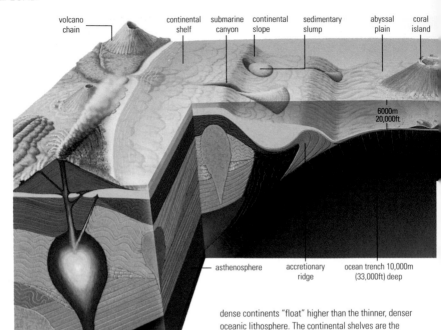

volcano chain — continental shelf — submarine canyon — continental slope — sedimentary slump — abyssal plain — coral island

6000m
20,000ft

— asthenosphere   accretionary ridge   ocean trench 10,000m (33,000ft) deep

▲ **lithosphere** The lithosphere "floats" in the Earth's mantle, and this defines where the surface of the crust – oceanic or continental – will be. Thus, thicker, less dense continents "float" higher than the thinner, denser oceanic lithosphere. The continental shelves are the transition. Chains of volcanoes or volcanic islands are often found at active margins, as shown in the – vertically exaggerated – diagram. The sediment that is scraped back when the oceanic plate subducts beneath

load The sediment carried by a river, ice or the sea. The load will vary depending on the amount of available material, as well as the carrying power of the transporting agent. Large glaciers and ice sheets can carry a greater load than a thin glacier, and will dump greater amounts of **till** when the ice melts. A river's load is greatest during times of flood, and quite small streams, which normally only carry sand and mud, can move large boulders on occasions. The load may be in solution, suspension, or rolled along the bed. The maximum load will be determined by the volume and velocity of the river, and if the full load is being transported any additional material will result in some deposition. In times of flood, some of the load will be deposited across the **flood plain**.

loam A type of soil which is between a clay and sand in characteristics. It has the good features of both clay and sand, and is often rich in humus. It is normally a rich and productive soil.

local climate The weather conditions experienced in a small area, such as a valley or a city. Local climate can also include studies of very small areas, but strictly this should be called **microclimatology**. Micro studies consider variations within an area of 0.4 sq miles [1 sq km] or less, or changes below the height of the **Stevenson screen**.

location quotient A method for measuring the concentrations of a particular industry. Data for different regions has to be compared with information for the entire country. By use of the location quotient formula, it is possible to make a quantitative comparison between the different regions.

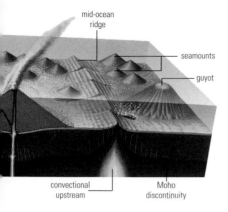

mid-ocean ridge

seamounts

guyot

convectional upstream

Moho discontinuity

the continental plate forms the accretionary wedge. Submarine canyons are more common on passive margins, but can be found here too. They were probably caused by surface erosion during periods of low sea level, and were subsequently widened by turbidity currents, or "submarine avalanches" of sediment. Many parts of the seafloor are highly fractured, especially around the mid-ocean ridge. There are many volcanoes; those that never emerge are seamounts. Guyots are volcanoes with their tops cut off flat, probably by wave erosion when the sea level was lower. Their great weight has sunk them down into the ocean crust.

location theory A theory or idea which tries to generalize and simplify the location of various patterns of land use. The **Von Thunen** model for agriculture is one such example, since it locates, in theoretical terms, the most intensive types of farming next to the farm. The least intensive types of farming are located at the greatest distances from the farm, and the economic reasons for this can be explained quite simply. The **Weber** industrial model is another example.

loch In Scotland, a loch is a lake, but in Ireland it is an arm of the sea, usually narrow and bounded by steep sides.

lode A vein or seam of minerals. Often several different minerals occur along the same lode. Lodes were probably formed as a result of liquids or gases, heated by vol-canic activity, forcing their way through existing rocks, and then cooling and solidifying. Some gold deposits in Australia and Alaska, USA, were found in lodes.

loess Wind-blown soil which consists of very fine particles of dust. It is generally very fertile, like a **loam**, and is often very **porous**. Much loess is formed by winds blowing dust from deserts. The largest expanse is in northern China in the valley of the River Huang He (Hwang Ho), where dust from the Gobi Desert has accumulated. As it is very soft, erosion can be quite rapid and the Huang is sometimes called the Yellow River because of all the loess it transports. Erosion wears many small gorges in the loess, and local people sometimes live in caves dug into the walls of the valleys. There are several areas of loess in northern France, where it is called "limon." Winds blowing from the ice-cap which lay further north during the last glacial phase probably picked up fine particles that had been deposited by streams of meltwater.

longitude The angular distance of a point on the Earth's surface measured east or west from a prime meridian. Since 1884, the prime meridian of longitude has been located at Greenwich in London, England. Longitude is measured in degrees east or west of Greenwich, which is 0°. All lines of longitude are **great circles**, and they extend from the North Pole to the South Pole, intersecting the Equator. As they all merge at each pole, they are obviously not parallel. A precise location of a place can be given by quoting its longitude and latitude coordinates. Time zones are related to the lines of longitude, and 15° of longitude represents a time change of one hour.

longitudinal coast SEE **concordant coast**

long profile A longitudinal section of a geographical feature. A long profile is especially useful to show the shape of a river valley, from its source to the sea; it is a smooth and regular curve, except where broken by waterfalls, in contrast to the

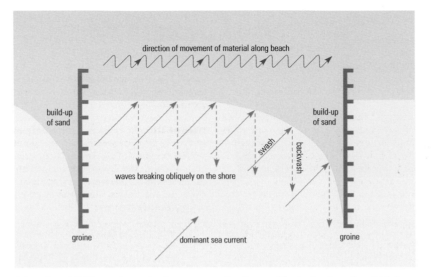

direction of movement of material along beach

build-up of sand

build-up of sand

swash

backwash

waves breaking obliquely on the shore

groine

dominant sea current

groine

▲ **longshore drift** Waves arrive at a beach at an angle influenced by the dominant sea current, whereas the backwash is at right angles to the beach, with water moving down the slope of the beach. The overall effect is to move material along the beach. Groines protect the coastline by reducing the distance material can travel as a result of longshore drift.

irregular **long profile** of a glacial valley. ALSO CALLED **talweg, thalweg**

longshore drift The movement of sand, mud and shingle along a coastline. The direction of drift can change, as it is caused by waves, which in turn are caused by wind. Most coastlines have a prevailing direction for longshore drift. Depositional material is moved up shore by the swash, which is at right angles to the crest of the waves. The material rolls back down

the shore with the **backwash**, which is at right angles to the coastline. In this way, particles are moved along the coast by successive wave movements. In order to reduce longshore drift, many cities have now constructed **groines** at right angles to the shore. Sand piles up on one side of the groine, but there will be little or no deposition on the leeward side of the groine.

lough In Ireland, a **loch** or **lake**.

low A **depression** or region of atmospheric pressure. In a low, the atmospheric pressure is lower than in the surrounding areas, and it will be a region of ascending air and probably **cloud**.

low-order goods SEE **convenience goods**

M

macroclimate The general climate of a large area; for example, for one particular region. SEE ALSO **microclimatology**

magma Molten rock beneath the surface of the Earth which forms **igneous** rocks when it solidifies. Below the surface, cooling is slow and large crystals form as the rock solidifies. Large masses of magma are in **batholiths**, and thin seams of magma create **sills** and **dykes**. If magma reaches the surface, it flows out as lava. The differ-ent igneous rocks formed by magma vary according to their chemical content as well as the depth of formation. SEE ALSO **extrusive rocks**

magnetism The Earth can be compared to a huge bipolar magnet. Compasses point toward the North Pole or the South Pole, depending on which hemisphere you are in. The magnetic poles are not quite located at the true poles, and their locations change slightly each year.

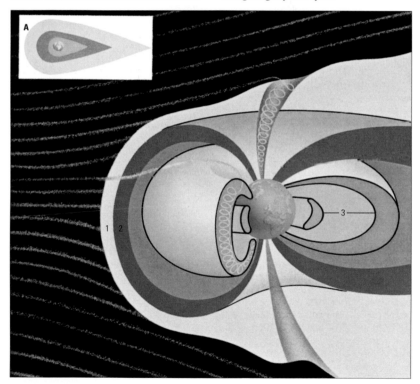

▲ **The magnetosphere** is the region in which the Earth's magnetic field can be detected. It would be symmetrical were it not for electrically charged particles streaming from the Sun (A), which distort it to a teardrop shape. The particles meet the Earth's magnetic field at the shock front (1). Behind this is a region of turbulence, and inside the turbulent region is the magnetopause (2), the boundary of the magnetic field. The Van Allen belts (3) are two zones of high radiation in the magnetopause. The inner belt consists of high-energy particles produced by cosmic rays and the outer belt of solar electrons.

maize SEE **corn**

malaria A widespread disease, found especially in tropical latitudes. It is transmitted by the *Anopheles* mosquito, which breeds in stagnant water. It is proving difficult and expensive to eliminate malaria, but there are now medicines which can reduce the risk of catching the disease. Malaria used to occur in many more parts of the world, including the USA in southern California and around the Mississippi River, and in southern Europe. In recent years a more resilient type of mosquito has been extending the range of malaria across northern Africa.

mallee A type of scrub consisting mainly of low-growing *Eucalyptus* and occurring in Australia, especially in areas with a semi-arid or dry **Mediterranean climate**.

malnutrition Inadequate nutrition or nourishment. Malnutrition is not the same as "undernourishment," which is simply an insufficiency of food. Malnutrition is generally associated with a lack of vitamins and some essential minerals, and always a shortage of protein. It leads to poor health and a lack of energy, which often means that less money can be earned or less food grown. Several ailments result from malnutrition, including beriberi and kwashiorkor. Children are most seriously affected by malnutrition, and there are high infant mortality rates in several parts of the world, but especially in Africa, because of this problem.

Malthus, Thomas Robert (1766–1834) An English political economist, who studied the growth of population. In 1798, he wrote down his conclusions; he believed that the population was beginning to increase at a geometric rate and would double in 25 years. At the same time, agricultural productivity in the best farming areas could only increase food supplies at an arithmetic rate. This, he forecast, would lead to overpopulation and an insufficient supply of food. Malthus suggested that positive steps should be taken to overcome the problem. He advocated later marriages and observed that famine, disease and war would all check growth. His gloomy forecast was proved incorrect in the short term because of changes which he had not foreseen, but the problems he described have been relevant in the **Third World** during the 1970s and 1980s.

mangrove Any tree or shrub of the genus *Rhizophora,* characterized by the ability to survive in salt water. Mangroves grow on the shore. They produce masses of aerial roots, some of which anchor themselves in the mud on the seabed and send up new plants, thus extending the mangrove colony in a seaward direction. As the trees become denser, sediment is trapped around the roots and a swamp is created. Gradually, the mangrove swamp becomes drier and firmer, as more muddy sediment accumulates. The drier parts will become inhabited by other plant species, while the mangroves continue to extend into new areas. Mangroves are found in tropical areas; for example, Malaysia and Indonesia.

man-made fiber SEE **synthetic fiber**

mantle 1 A structural zone of the Earth, situated between the **lithosphere** and the **core**. It is separated from the lithosphere by the "Moho." The mantle is at an average depth of 20 miles [30 km] beneath the land, but only 6 miles [10 km] beneath the oceans. The rocks of the mantle are **ultrabasic**; however, variations in seismic velocities indicate that the upper mantle is less homogenous than the lower mantle. SEE ALSO **Mohorovičić discontinuity**. **2** The surface accumulation of soil and weathered rock fragments.

manufacturing industry The production, now usually a mechanized process and on a large scale, of finished or partly finished products from raw materials. For example, a clothing factory uses wool, cotton, etc., to produce clothes.

map projection A system for representing on a flat surface the curved surface of the Earth, using a grid which corresponds to the lines of **latitude** and **longitude**. As a globe is round, and maps are flat, there is inevitably a certain amount of distortion when maps are drawn. Some projections give true shape or equal area; others may be based on the Equator and will become less accurate toward the poles. They can be divided into three broad categories – conic, cylindrical and azimuthal. **Mercator's projection** is an example of a modified cylindrical projection and is fairly accurate, although it exaggerates the size in high latitudes. The Mollweide projection is also a modified cylindrical projection. An example of a conical projection is the Bonne projection: here true scale along the meridians is sacrificed to enable the accurate representation of areas. Cartographers will use different projections depending on the map scale, the size of the area to be mapped, and what they want the map to show.

maquis A type of scrub vegetation found in Mediterranean regions of France. It consists of low evergreen bushes and small trees, including myrtle, laurel, olive and arbutus, which have the capacity to survive the hot dry summers by means of water-conserving devices, such as narrow or waxy leaves. They also have long roots to reach underground water supplies. Many of the plants are aromatic. During World War II, the French resistance movement adopted the name "maquis," because it was always possible to hide in this fairly dense evergreen scrub vegetation. It is known as "macchia" in Italy, and is similar to the **chaparral** in California and **mallee** in Australia.

marble Limestone which has been metamorphosed by the effects of heat or pressure. The decorative and patterned effects in marble may be the result of the fossil content of the original limestone. Most marbles are light or whitish in color, but the presence of small quantities of different minerals can produce a variety of other colors. Marble can be polished, which gives it much of its commercial value. Although there are limestones which can be cut and polished, they are generally not as decorative as marbles.

MAP PROJECTIONS

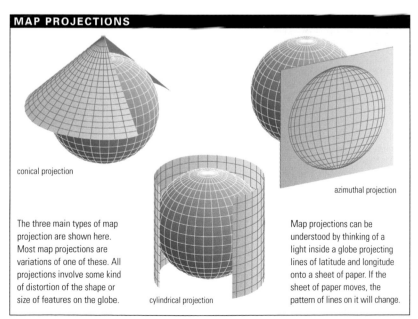

conical projection

azimuthal projection

The three main types of map projection are shown here. Most map projections are variations of one of these. All projections involve some kind of distortion of the shape or size of features on the globe.

cylindrical projection

Map projections can be understood by thinking of a light inside a globe projecting lines of latitude and longitude onto a sheet of paper. If the sheet of paper moves, the pattern of lines on it will change.

marginal land An area of land which is not very productive and needs to be farmed on an extensive basis. If economic returns are not good, it may not be worth farming at all. Marginal land is often on hillsides. To make the land more productive, money would have to be spent on fertilizers, etc., and the cash returns may not be sufficient to make this worthwhile. There are also marginal areas of semiarid grasslands close to the wetter and more productive grasslands, such as the **prairies**. Government grants are sometimes available to farmers in marginal areas as an encouragement to keep them in the area.

maritime air mass A large mass of air which has come from an area of sea. A tropical air mass often comes from the Gulf of Mexico before flowing over the USA, whereas a polar air mass will have come from the Arctic. The weather associated with the two air masses will be very different.

maritime climate A climate which is strongly influenced by proximity to the sea. The oceans warm up much more slowly than the land, because the heat of the sun is spread out through a great depth of water and ocean currents allow the heat to move vertically as well as horizontally. However, they also retain heat for much longer than the land. For this reason, climatic conditions near the ocean tend to be much warmer in winter and slightly cooler in summer than in inland areas at the same latitudes. In addition to the effects on temperature, the sea also influences precipitation, which is greater than in inland locations. Maritime climates are found in British Columbia in Canada, Washington and Oregon in the USA, Britain, Norway, northwest France, southern Chile, Tasmania, and South Island in New Zealand.

market area The catchment area for a central place. It is the area from which customers will travel in order to obtain goods and services. The size of the market area is determined by the size of the central place. Theoretically, market areas should be hexagonal in shape, but the real world is not quite as simple as the Christaller model. SEE ALSO **central place theory**

market gardening SEE **horticulture**

market town British name for a town where a market is held regularly or permanently. Traditionally in England, permission to hold a market had to be granted by the monarch to a town.

marl A type of clay which contains large quantities of calcium carbonate and generally forms a productive alkaline soil.

marsh A soft wet area which suffers from poor drainage and frequent waterlogging. Many marshy areas are completely flooded for part of the year, becoming a little drier periodically. They are generally associated with areas of **impermeable** rock, and are often flat and low-lying areas of land, where the water table is near the surface. Some marshy areas are also found in hill regions where rainfall is very heavy and **peat** bogs form. It is generally possible to drain marshes, by digging deep ditches and lowering the water table. Draining changes the **ecosystem** completely and wildlife habitats can be destroyed. Conservationists normally wish to leave marshlands in their present state, while farmers often wish to drain them in order to make them productive.

massif A distinct area of higher ground or mountain with fairly similar features. Good examples are are the Massif Central in France and Vinson Massif, Antarctica

mass movement The downhill movement of soil and rock due to the effects of gravity. Water can act as a lubricant, increasing movement after heavy rainfall or a period of snow melt. **Freeze-thaw** activity helps to provide loose material, which accumulates on sloping ground and then gradually slides downhill. There are various types of mass movement. Slow movements are called soil creep or rock creep. Faster movements are called earth flow or earth slide, and rock fall or rock slide, and these may be more dramatic and visible to the naked eye. ALSO CALLED **mass wasting**. SEE ALSO **creep, soil creep, solifluxion**

mass production The large-scale production of manufactured goods, often using an automated factory and a conveyor belt system. Machinery creating a large number of identical products can usually produce goods more cheaply than in small factories.

mass wasting SEE **mass movement**

maximum thermometer A thermometer which records the highest temperature of the day. Usually found together with a minimum thermometer, it contains a small metal rod above the column of mercury. As the temperature rises, the mercury moves up the tube, pushing up the metal rod. When the temperature starts to fall, the metal rod is left at the highest point which has been reached. It is then possible

to read off the highest temperature later in the day or early the following morning, which is the usual time for taking temperature readings. By means of a magnet, it is possible to move the metal rod along to the mercury, in order to reset it. SEE ALSO **minimum thermometer**

meander A bend or curve in the course of a river. A river will wind around any obstacle, such as hard rock, or even a pebble. Once a meander has been created it will continue, becoming accentuated by the erosive action of the river. On the outside of a bend there will be lateral corrasion, which will gradually work out sideways. Meanders gradually move downstream, too. On the inside of the bend there is likely to be some deposition, which will build up a flat **flood plain**. Meanders tend to be quite small in the more mountainous upper course of a river, but they become larger further downstream. Where spurs still remain, there will be **interlocking**

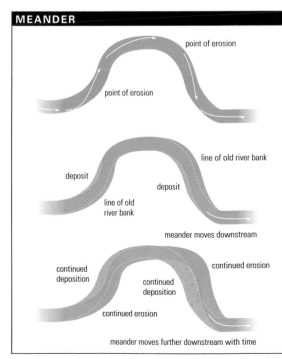

MEANDER

point of erosion

point of erosion

line of old river bank

deposit

deposit

line of old river bank

meander moves downstream

continued deposition

continued erosion

continued deposition

continued erosion

meander moves further downstream with time

A meander will gradually move downstream as a result of a continuing process of deposition and erosion. The current in a river tends to erode the outer side of any bend more quickly than the inner side. At the same time, material will tend to be deposited on the inner side of curves. Over time, the course of a meander may move until not part of it follows the original course.

spurs, around which the river will meander. Down in the lower course, the meanders may spread over very large distances, and some on the Mississippi River in the USA are up to 40 miles [60 km] in length. The erosion on the outside of the bend, especially in the upper and middle reaches, will form a steep bluff and will gradually widen the valley. On the inside of the bend will be the slip-off slope, which is an accumulation of sediment. Meanders may eventually give rise to **oxbow lakes**. The term is derived from the name of a winding river (now Mendere) in Turkey. SEE ALSO **lateral erosion**

mean sea level The average level of the sea. Precise measurements are collected at various locations around the world in order to arrive at an official height. Many other measurements are based on mean sea level. SEE ALSO **ordnance datum**

mechanical weathering SEE **weathering**

medial moraine SEE **moraine**

median In statistics, the value found in the center of a set of data ranked in size order. The median has the same number of values above and below it, and if there is an even number of values it is placed halfway between the two central values. The "upper quartile" covers the highest 25% of the values and the "lower quartile" covers the lowest 25%. SEE ALSO **quartile**

Mediterranean climate A type of climate associated with the countries surrounding the Mediterranean Sea, but also found in southern California in the USA, central Chile, near Cape Town in South Africa and near Perth and Adelaide in Australia. Summers are hot and sunny, averaging 80°F [25°C] or more; winters are mild, ranging from 40°F to 60°F [5°C to 15°C]. Summers are very dry, but winters can be quite wet, with up to 20 in to 25 in [500 mm or 600 mm] of rainfall. Some of this rain comes as **thunderstorms** at the end of the summer.

megalith A large stone, erected by humans, often to make a monument. Megaliths may be arranged in rows or circles and have been very important to past cultures, many dating from between about 3000 BC and 2000 BC. Examples can be seen at Stonehenge in Wiltshire, England, and Carnac on Quiberon Bay, France.

megalopolis A very large city. The term is often used to refer to the area in the northeast USA that stretches from Boston to Washington, D.C., and includes New York, Philadelphia and Baltimore. It can also be used to describe the area in California that stretches from San Francisco to Los Angeles and San Diego, Honshu in Japan, and various locations in Western Europe.

meltwater Water created from the melting of snow or ice. This usually happens in spring as temperatures begin to rise and can cause flooding. Glacial meltwater can cause a rise in river levels later in the year or may form its own channel in a formerly glaciated area.

mental map A map formed in a person's mind through his or her knowledge and prejudices; it will reflect that person's perceptions of a particular area or place and can therefore be very useful and interesting to observe how different people in different areas view places and people around them.

Mercalli scale A scale for measuring the intensity of earthquakes, numbered from I to XII; I indicates no damage and no real visible effects, although the seismograph records a slight tremor, while at XII on this scale there is total destruction and devastation, including the collapse of large buildings. It has now been replaced by the **Richter scale**.

Mercator's projection A popular type of map projection, often used for showing the whole world. It is a rectangular map, based on a cylindrical projection. It was first used by Gerhard Mercator in

1569. All lines of latitude are the same length as the Equator, whereas on a globe they become shorter toward the poles. At 60°N, the line of latitude is only half the length of the Equator on the globe. To balance this doubling of scale on the projection map, there is also a north-south doubling. This preserves accuracy of shape, which is one of Mercator's major advantages. However, there is considerable distortion in the high latitudes, which means that Greenland looks much larger than Australia. The equal stretching of the land in all directions means that directions are accurately portrayed, and so the Mercator map is good for showing routes.

mercury barometer SEE **barometer**

merry-go-round A system of transportation in which trains or trucks perform the same journey repeatedly, sometimes on a circular route. It occurs commonly at power stations where there are direct links to a coal field; a locomotive with a line of cars goes through the loading hoppers at the coal mine and then moves to the power station, where it unloads onto the stockpile of fuel, before going back to the coal mine to load up again.

mesa A flat area of upland rather like a table, so called after the Spanish *mesa*, meaning "table." Mesas are commonly found in arid or semiarid areas, such as Arizona, Utah and New Mexico in the southwest USA. They often have horizontal strata of **sedimentary rock**, and their location in dry regions is also significant. The lack of rainfall means that rivers are the major source of erosion. But they can cut deep valleys, which remain steep-sided, partly because of the horizontal structure, but also because there is little rainwash to wear away the sides and change the cross-profile from a U-shape to a V-shape.

meseta Similar to a **mesa**, this term refers specifically to the high plateau area in central Spain, which tilts toward Portugal.

mesopause The boundary between the **mesosphere** and the **thermosphere**, at

▼ **mesa** Mesas are most usually found in arid or semiarid areas with horizontally stratified sedimentary rock. They are often associated with canyons and buttes, which form in similar conditions. All of these features are produced by river erosion cutting channels into the landscape.

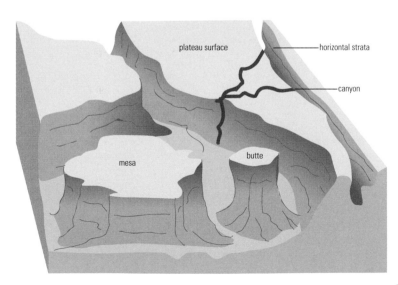

about 50 miles [80 km] above the Earth's surface. SEE ALSO **atmosphere**

mesosphere The zone of **atmosphere** which extends from the **stratopause** (at about 30 miles [50 km] above the Earth's surface) to the **mesopause**.

Mesozoic The era of middle life, in the **geological time scale**. The Mesozoic falls between the **Paleozoic** and **Cainozoic** eras. Whereas the Paleozoic was the time of primitive life, the Mesozoic was the time of reptiles. It lasted from about 225 to 70 million years ago, and contains the Triassic, **Jurassic** and Cretaceous periods. SEE ALSO **geological column**

mestizo A person of mixed European and American Indian ancestry. When the first Europeans arrived in Latin America, there were virtually no women among the early settlers; this led to intermarriage with the Indians. The early mixing of the races in Brazil and other countries has been a major reason for the relative lack of racial problems in South America.

metamorphic rocks Rocks which were originally **igneous** or **sedimentary**, but have been changed by the effects of heat or pressure, or both heat and pressure. Heat is the result of volcanic activity, and pressure is the result of **earth movements**.

Metamorphic rocks are generally hard and resistant to erosion, and are likely to form high ground. In metamorphic rocks, minerals may recrystallize or be compressed, and the new rock may look completely different from the original sedimentary or igneous rock. Limestone becomes **marble** when metamorphosed; other changes include shale to **slate**, sandstone to **quartzite**, and granite to **gneiss**. Examples of metamorphic rocks can be found in large areas of Canada, the USA, Finland and Western Australia.

métayage SEE **sharecropping**

meteorology The study of the **atmosphere**. This includes pressure, temperature, clouds, winds, etc., and the description of weather and its explanation.

mica A rock-forming mineral made up of silicates, which is characterized by a platy habit, easily splitting into thin sheets.

microclimatology The study of weather conditions in a small area and a short

▼ **metamorphic rock** Some common metamorphic rocks are shown below: igneous rock (1) may become gneiss or schist; sandstone (2) may become quartzite; shale (3) may become slate, and if the metamorphism is more pronounced, gneiss; limestone (4) may become marble.

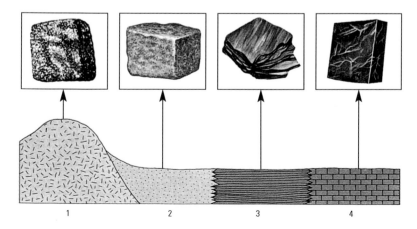

distance above the ground. The size of area studied may range from a few square feet to less than 0.4 sq miles [1 sq km]. However, the study of urban climates may cover an area the size of London or Paris, but would still be considered a micro study. Microclimatology is sometimes defined as the weather below the height of the **Stevenson screen**. Variations between the weather at ground level and conditions a 3 feet [1 meter] or so higher can be very important to plant growth. Exposure to wind, or shelter from it, can also be significant. SEE ALSO **local climate**

Mid-Atlantic Ridge The ocean ridge that runs for 8,700 miles [14,000 km] along the middle of the Atlantic Ocean. It is formed at a constructive margin, where two plates are diverging and new magma is flowing out to form new ocean crust. Iceland is located on the ridge itself and was formed by the outpourings of volcanic lava. SEE ALSO **plate tectonics**

migration The movement of people (or animals) from one place to another. Migration may be permanent or temporary, and may be within one country, or from one country to another. People migrate in order to get away from something they do not like (push factor), or to go to somewhere which seems more attractive (pull factor). It will often be a combination of "push" and "pull" which makes people migrate. In Britain, there has been migration to London and the southeast for many years, although there is now a slight reversal of this trend; in Italy, the movement has been from south to north. Major migrations in the past have included the movements from Britain and Western Europe to North America, Australia and New Zealand. More recently, many Jewish people have migrated to Israel from Russia and elsewhere, and workers from southern and Eastern Europe have migrated to Western Europe. Mass migrations of people from rural areas to the big cities have become a serious problem for many poorer countries, where **shanty towns** have sprung up around most large cities to house the increased population.

millibar A unit of atmospheric pressure which is equal to one thousandth of a bar. It is used internationally on weather maps, on which isobars are drawn at 2 mb or 4 mb intervals. Average pressure in temperate latitudes is about 1,012 mb; 1,000 mb is equal to 750 mm [29.53 in] of mercury.

millionaire city A city which contains more than 1 million inhabitants. The number of millionaire cities has risen rapidly since 1960, and by 1994 there were over 300. A large number of them are located in lowland regions, often near the coast, and many of them are capital cities.

millstone grit 1 A type of coarse sandstone. It is best seen in the Pennines in England, where it forms high, rather bleak moorland. The rock was named "millstone grit" because it was used to make grindstones for mills. **2** The geological age, part of the **Carboniferous** period between the Carboniferous limestone and the **coal measures**, during which millstone grit was formed. In the Millstone Grit period, there were other rocks in addition to the coarse sandstone; fine sandstones, shales and thin limestones, as well as thin coal seams can also be seen. They were all caused by fluctuations of sea level at the time when the rocks were accumulating, about 320 million years ago.

mineral An inorganic substance with a particular chemical composition. Many minerals have a distinctive crystal shape; for example, cubic fluorspar or **pyrites**. Rocks are combinations of minerals.

minifundio Very small farms in Latin America, often farmed by peasants.

minimum thermometer A thermometer which records the lowest temperature reached during the night. It does this by means of a metal rod which is placed inside the U-shaped tube with the mercury or alcohol. As the temperature falls during the night, the mercury or alcohol moves along the tube, pushing the metal rod ahead. As the temperature warms up again in the morning, the mercury moves back along the tube, leaving the metal rod at the coldest point, so that the minimum temperature can be read off. The metal rod can be moved along the tube with the help of a magnet. The minimum thermometer is usually found together with a **maximum thermometer**.

misfit A stream that is too small for the size of the valley in which it is situated. This may be the result of **river capture**, in which a river has been beheaded, and so is much smaller than it used to be. Many misfits can be seen in glacial **U-shaped valleys**, where the ice has made the valley deeper and wider than would normally be associated with the size of the river.

mist A cloudlike mass of minute droplets of water suspended in the air near the ground. When air is cooled, it cannot hold as much water vapor. As the water vapor condenses, the drops become visible, forming mist. Mist is like thin fog, with a visibility of at least 3,000 ft [1,000 m]. It is most likely to occur in calm, settled anticyclonic conditions, especially in the fall. It is generally very patchy, occurs most frequently near lakes, rivers and coastal areas where there is more water available, and tends to accumulate in hollows. Once a wind starts to blow, turbulence will mix the air, and the mist will clear. In **urban** areas, where there may be pollution, the mist may become dark and dirty, to form **smog**.

mistral A cold northerly wind which blows down the Rhône valley in France. It is caused by cold air from Europe blowing southward into a low-pressure area over the Mediterranean. The depressions occasionally pass along the Mediterranean from the Atlantic, especially during the colder half of the year. When the mistral blows down the Rhône, temperatures fall by several degrees, and the strong winds of up to 40 mph [60 km/h] make conditions very unpleasant. Most farmhouses have rows of trees on their northern side as protection from the wind, and many fields of fruits and vegetables also shelter behind

rows of cypress trees, which always extend from east to west and are very numerous in parts of the Rhône valley. In the Camargue area of the Rhône delta, the landscape is very flat and windswept. When the mistral blows, temperatures sometimes fall as low as freezing point, and many of the houses have doors and windows on the south-facing side only. A similar wind is experienced in other locations along the Mediterranean coast of southern Europe.

mixed farming Combined arable and pastoral farming. In the past, many farmers would rear animals and grow crops. The animals provided manure for the fields, which helped to maintain fertility. With crop rotation as well as animal manure, the structure of the soil could be maintained easily. In order to increase yields, many farmers turned to **monoculture** and had to spend large sums of money on fertilizers, which, it is now realized, can do great harm to the structure of the soil. Short-term gains in profit could be followed by long-term devastation or destruction of the soil. Because of this, several farmers are now reverting to mixed farming, with crop rotation, pasture and animals all forming part of the farming activity.

model A representation, usually on a smaller scale, of some feature in the real world. Models are used to help understand the workings of such features and the relationships between different ones, in a simplified and generalized way. An iconic model shows a scaled-down version of reality, using real objects; an analog model uses diagrams or maps; and a symbolic model uses mathematical symbols and equations. It is difficult to represent human behavior in models. Examples include the **von Thunen** land-use model, and the **Burgess model**.

Moho SEE **Mohorovičić discontinuity**

Mohorovičić discontinuity The boundary between the **lithosphere** and **mantle**, named after A. Mohorovičić who discovered it in 1909. The depth varies from place to place, but it is much deeper beneath the land masses than under the sea. The changing density of rocks at the "Moho" causes earthquake waves to change speed, from about 6.4 km/sec [23,000 ft/sec] above the Moho, to about 8.2 km/sec [27,000 ft/sec] in the rocks below. ALSO CALLED **Moho**

Mohs' scale A scale of mineral hardness, named after its inventor F. Mohs (1773–1839), a German-Austrian mineralogist. The scale is from one to ten:

1   Talc – easily scratched by a fingernail
2   Gypsum
3   Calcite
4   Fluorite
5   Apatite
6   Feldspar – the hardness of steel
7   Quartz – the hardest of the common minerals
8   Topaz – semiprecious stone
9   Corundum – used in carborundum stones for sharpening knives
10 Diamond

monadnock An isolated hill, possibly the last remnant of a large range of hills which have been eroded for millions of years. A monadnock is a **residual hill**; it is probably a harder rock than the surrounding rocks, and so they have been eroded to a lower level. The name is derived from Mt. Monadnock in New Hampshire, USA, and there are many monadnocks in the southwest USA, where residual hills are often seen in Western movies.

monocline A type of fold in which one limb (slope) is vertical or nearly vertical for a short distance, though the general trend of the limb of the fold may be quite gentle.

monoculture The growing of one crop. Monoculture has been a common practice for a long time in many places, such as plantations or vineyards, but it has become far more widespread during the last 20 or 30 years, especially in the production of cereals. This is as a result of the availability of machinery for harvesting large quantities very quickly. There are also many fertilizers which restore fertility to the soil, so that

crop rotation is unnecessary. Monoculture has created problems, however, as dependence on one crop can be serious if the price suddenly falls because of a glut, or if new pests invade the crop. Moreover, it is being realized that monoculture harms and possibly ruins the soil, in spite of the addition of artificial fertilizers. Monoculture in such areas as the prairies in Canada and the USA, the steppes in Russia, and East Anglia in the UK is now declining slightly, and there are signs of a return to crop rotation methods. SEE ALSO **rotation of crops**

monopoly When a particular good or service is provided by a single supplier. This usually happens when one firm in a market controls most of the output of an industry. Governments in capitalist countries may try to encourage fewer monopolies and more choice in a market, since competition may improve the quality and price of the final product.

monsoon A seasonal wind, especially in southern Asia. In summer, which is the monsoon season, the winds normally blow from the sea to the land and bring rain, but in winter there is a complete change of direction and the winds blow out from the land, giving dry weather. Some monsoon regions are very wet. Cherrapunji in India, for example, receives over 430 in [11,000 mm] per annum, but others can be dry, such as the Thar Desert between India and Pakistan where the rainfall is less than 10 in [250 mm]. The major monsoon areas are in Asia, where the seasonal reversal of wind is greatest. This is because the largest continent, Asia, is adjacent to the largest ocean, the Pacific. In the smaller continents of South America, Africa, Australia and North America, the monsoonal effects are less marked. These smaller continents do not have such wet summers or such dry winters, and are sometimes referred to as "eastern marginal" rather than true monsoon. The rainfall brought by the monsoon conditions in Asia is cyclonic, but there is the addition of **orographic rainfall** in the mountains, which for this reason are often much wetter than the lowlands. In addi-

▲ **monsoon** Caused by low pressure areas over land masses in the summer, monsoon winds bring wet air from the sea and so heavy rain. In the winter, high pressure areas cause dry winds to blow from the land. During the summer (left) large areas of mainland Asia are heated by the Sun. The air over these regions expands and rises forming regions of low pressure.

Wet winds from the sea then blow into these areas giving the summer monsoon season. In the winter (right), the situation is reversed and regions of high pressure are formed over the land, which is relatively cool. Dry winter monsoon winds then blow out to sea. The paths of these winds are deflected due to the Coriolis effect.

tion, the great heat during the day often gives rise to **thunderstorms,** and convection rainfall occurs.

montane Meaning mountainous, or of a mountain area, this can be used to describe the montane "cloud" forest found in upland tropical and equatorial regions.

moor A tract of bleak open land, generally on a plateau or mountain. The vegetation is likely to be heather or bracken, with some coarse grasses. Many moorland areas have extensive stretches of **peat** bog; for example, parts of the Pennines and Dartmoor in England. Acid soils often develop, especially in areas of heavy precipitation.

moraine An accumulation of boulders and rock fragments which have been deposited by ice. Some of the rocks may have been eroded by the ice; some may be the result of **freeze-thaw** activity; and others are weathered blocks that simply have been transported by the ice. Morainic debris is

carried by the ice and dumped at the sides or at the end, or at the bottom of the ice if it falls through the crevasses. Accumulations at the side of a glacier are called **lateral moraines**; at the end of the glacier they are called **terminal moraines.** When two glaciers merge, two lateral moraines unite in the middle of the enlarged glacier to form a "medial moraine." Beneath the ice there will be gradual accumulations of ground moraine, but much of it is reduced to small fragments and powder by the effect of the ice grinding it on the **bedrock.** Ground moraine can become an effective abrasive tool, if scraped along the valley floor by the movement of the ice. Rock particles actually within the ice are described as "englacial," but when the ice all melts, they fall to the floor and become part of the ground moraine.

morphology SEE **geomorphology, urban morphology**

mountain A mass of land that rises fairly steeply from the Earth's surface to a summit. In Britain, the term is applied to a mass of over 1,000 ft [300 m] in height; beneath this height, the term "hill" is used. The 30 highest mountains in the world are all in Asia.

Mozambique Current A warm **ocean current** found off the coast of south-eastern Africa. This affects the climate of the nearby countries, so, for example, Mozambique has a tropical climate with **coral reefs** lying just offshore. ALSO CALLED **Agulhas Current**

mudflow The movement of a mass of soil or rocks, liquefied by rain or melting snow.

multinational A large company which has interests in several countries. Many multinationals have become very influential in one particular product, and may control production and marketing in certain countries. This may be beneficial, but there are certainly situations in which a multinational may take a course of action which benefits the company but does not really help the local people or the country. It is a problem which affects **Third World** countries in particular. Tea and bananas are produced by multinational companies, which can influence developments in the growing areas. They are often accused of using cheap labor and exploiting the local people; on the other hand, they claim that they are helping development in the area by providing jobs, and also schools and hospitals on occasion.

multiplier effect The way in which the creation of one or two industries may encourage the setting up of other industries in the same area. For example, a steel works may provide raw materials for an engineering works, which may provide machines for a textiles factory, and so on. Once started, the effect will continue, because there will be a need for transportation, schools and hospitals, which will provide more jobs and also create a good local market. More people will be encouraged to move into the area, and possibly additional industries will be attracted.

WORLD'S HIGHEST MOUNTAINS (by order of size)			
Name	Location	Height m	ft
Everest	China/Nepal	8,850	29,035
K2 (Godwin Austen)	China/Kashmir	8,611	28,251
Kanchenjunga	India/Nepal	8,598	28,208
Lhotse	China/Nepal	8,516	27,939
Makalu	China/Nepal	8,481	27,824
Cho Oyu	China/Nepal	8,201	26,906
Dhaulagiri	Nepal	8,177	26,811
Manaslu	Nepal	8,156	26,758
Nanga Parbat	Kashmir	8,126	26,660
Annapurna	Nepal	8,078	26,502
Gasherbrum	China/Kashmir	8,068	26,469
Broad Peak	China/Kashmir	8,051	26,414
Xixabangma	China	8,012	26,286
Kangbachen	India/Nepal	7,902	25,925

Soon, administration will be necessary, providing yet more jobs.

multiple nuclei model One of the classical models of **urban morphology**, which shows that large cities develop and grow around two or more separate central places, or nuclei. The nuclei may have separate settlements which have been swallowed up by urban growth. **Harris and Ullman** wrote about this model in 1945, and they believed it added important ideas to the models described by **Burgess** and **Hoyt**.

mushroom rock SEE **pedestal rock**

muskeg A boggy area which contains sphagnum moss, and other mosses and lichens. It is found in the **tundra** regions of Canada and Alaska, USA, and also in the northern parts of the coniferous forest belt.

Myrdal, Gunnar A Swedish economist who, in 1957, wrote about his theory of **cumulative causation**. He believed that the developments of industry, agriculture and economic growth would occur most readily in areas where there were already successful activities. Success would breed success and attract other forms of development. His ideas fit in with those of John Friedmann, who formulated the "core-periphery idea," and in many countries attempts have been made to establish new core areas. SEE ALSO **core** [def. 1]

N

nappe SEE FOLDING

NATIONAL PARK A specially designated area of land, where development is controlled and planned. The world's first national park was Yellowstone National Park in the USA, established in 1872. National parks are usually left in a wild and natural state, and may contain many wild animals, high mountains, **geysers**, glaciers, etc. They generally cater for visitors, with hotels and camp sites, and provision for fishing and walking. Most national parks in the world aim to preserve or conserve the landscape and to enable the public to enjoy the countryside.

NATO SEE **North Atlantic Treaty Organization**

natural arch An archway formed by marine erosion. On an exposed headland, a cave may be enlarged until it cuts through to the other side of the rock to form a tunnel or an archway.

natural bridge A bridge of rock formed by erosion in a limestone area. When an underground stream erodes a tunnel, it forms a large cavern; if the roof then collapses, a gorge will result. In locations where two caverns have collapsed quite close to one another, a natural arch or bridge can be left behind; for example, Natural Bridge in Virginia, USA.

natural gas A combustible mixture of hydrocarbons in the form of ethane and methane found in trap rocks in the Earth's crust, often together with oil. The gas will need treatment or purification, but can be used as a very clean source of energy. Natural gas is an important source of energy in many countries, including the USA, Russia, the United Kingdom, Iran and the Netherlands.

natural increase The increase in population caused by the difference between the **birth rate** and the **death rate**. When the birth rate is 40 per thousand and the death rate is only 20 per thousand, the natural increase is very high. This is typical of many **Third World** countries. The overall position, however, depends on immigration (movement into an area) and emigration (movement away from an area) as well as natural increase. SEE ALSO **zero population growth**

naturalization 1 When a foreigner is given the citizenship of a country. **2** The introduction of an organism to a new habitat, where it can develop and reproduce.

natural resources Those materials used by man which are found naturally; for example, mineral deposits, energy, timber, fish, wildlife, the landscape. SEE ALSO **non-renewable resources**

natural selection That part of the theory of evolution developed by Charles Darwin based on "the survival of the fittest." Organisms reproduce and their "fittest" offspring will survive to pass on their best and strongest characteristics to the next generation. The weaker offspring will not survive in their environment and will therefore not pass on their disadvantageous characteristics.

natural vegetation The kind of vegetation which would grow naturally without interference from man. There are probably very few parts of the world where truly natural vegetation remains, as most areas have been cleared, even if there has then been a secondary growth of natural vegetation. Parts of the tropical forests and some areas in Arctic regions or on high mountains may still be as they were hundreds of years ago. However, vegetation does alter naturally.

nautical mile A unit of distance used in navigation. Slightly longer than a normal mile, it represents one minute of longitude, measured along the line of a **great circle** (60 minutes are equal to one degree of latitude or longitude of the Earth). The distance is about 6,080 ft [1,853 m].

neap tide A tide in which there is minimal variation between high tide and low tide as a result of the lunar and solar tidal influences pulling against one another. The opposite is a "spring tide." Spring tides are the result of lunar and solar tidal influences working together to create a high tidal range. Neap tides and spring tides happen every month, the spring tide occurring two weeks after the neap tide. Neaps are the result of the effects of the Sun and the Moon offsetting one another and balancing out their influences.

nearest neighbor analysis A technique for describing the distribution of features, such as churches, windmills, etc. The formula gives a precise quantitative answer, which will be between 0 and 2.15. The answer will show whether the distribution of the features is completely clustered (0), random (1) or regular (2.15 shows a uniform grid), and the figure for one area can be compared with that of another.

Neolithic The period of human development in the last stages of the **Stone Age**. Neolithic cultures developed at different times in different regions; for example, they had developed by around 1500 BC in Mexico and South America, about 4000 BC in Europe, and around 8000 BC in western Asia. More sophisticated techniques, such as the grinding and polishing of stone, were taken up at this time; it also saw the beginning of the domestication of animals and the planting of crops. SEE ALSO **megalith**

network An interconnecting pattern of roads, railroads, etc. It is often easier to draw communication and transportation networks as topological maps, which look neater and are often clearer; for example, a map of the New York Subway, London Underground or Paris Metro.

névé (*French*) SEE **firn**

newly industrializing countries Any of the less developed countries which have built up their industries in recent years. Outstanding examples are Taiwan and South Korea in the Far East. Brazil, Nigeria and many other countries have begun to use the wealth of their natural resources for industrial development. Because of low labor costs, they can sometimes undercut the old established industrial regions.

New Red Sandstone The Permian and Triassic geological periods. SEE ALSO **Paleozoic**

nick point SEE **knick point**

nimbostratus A low layer of dark clouds sometimes occurring at the warm front of a depression. SEE ALSO **cloud**

nitrogen cycle The circulation of nitrogen in the **ecosystem**. Nitrogen (a colorless gas which makes up some 78% of the **atmosphere**) is a vital constituent of living organisms and is used by man in the production of fertilizers and ammonia. In the cycle, atmospheric nitrogen is "fixed" (i.e. changed into organic compounds) by bacteria or blue-green **algae** (or sometimes by lightning strikes or industrial processes) and the resulting nitrates and ammonia are taken up by plants and animals, which change them into proteins and amino acids. These are returned to the soil through feces or the death and decay of the plants and animals, where "denitrifying" bacteria works on them to release free nitrogen back into the air.

nivation I Mechanical weathering caused by **freeze-thaw** activity. Water lying in the cracks in rocks freezes and expands. As the temperature rises, the ice melts, only to freeze again the following night. The repeated freezing and thawing, expanding and contracting, causes pressure and stress on the rocks, which eventually crack along any lines of weakness. Nivation causes the formation of angular fragments of rock, which accumulate at the foot of rocky outcrops. Accumulations on slopes form **scree**. Ex-

NITROGEN CYCLE

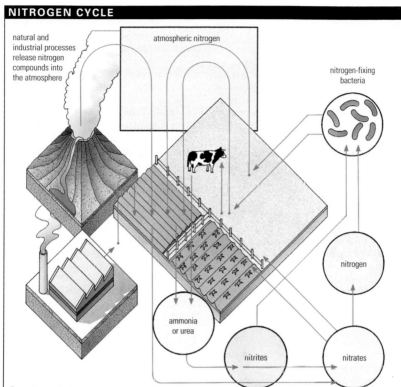

natural and industrial processes release nitrogen compounds into the atmosphere

atmospheric nitrogen

nitrogen-fixing bacteria

nitrogen

ammonia or urea

nitrites

nitrates

Free nitrogen (N_2) in the atmosphere cannot be absorbed directly by plants or animals. Soil-dwelling bacteria known as nitrogen-fixing bacteria convert the nitrogen into nitrates, in which form it can then be absorbed by plants. The plants are eaten by herbivores, the "primary consumers" of the standard food chain. Some of the excess nitrogen in the animals' bodies is converted to ammonia or urea and then returned to the soil. In addition, when the animals die, bacteria decompose the tissue and release any remaining nitrogen to the soil. The nitrogen cycle can then begin again.

amples can be seen in the Alps and Arctic environments. **2** The **freeze-thaw** activity which takes place around the edges of a snow patch. It may erode particles of rock to form a slight hollow, in which snow may accumulate again the following year. Nivation hollows may eventually become **cirques** if the freeze-thaw activity continues for many years.

node A point in a **network** where routes meet or intersect, such as a settlement or crossroad.

nomad A wanderer; a person with no fixed home. Most nomads nowadays are only seminomadic, as they have a home village in which they spend several months each year before going off on their wanderings. Some nomads wander in search of food for themselves, but most are pastoralists, who have to take their animals to different areas in order to find pasture. Nomadic pastoralism is a system of farming found in any different locations, including temperate grasslands, tropical grasslands, semideserts and tundra. In the **tundra** there are reindeer herders, such as the Lapps of northern Sweden and Finland; in central Asia there are cattle and sheep herders; and in the Sahara Desert there are camel herders. Just to the south of the Sahara is the **Sahel** region. Here there are many pastoralists, such as the Fulani. Most groups rear cattle, though some keep goats. They wander for several months each year seeking pasture for their cattle. There are now more people and more cattle than there used to be, and so much of the sparse pasture is being overgrazed. This has led to **desertification**, and many areas which used to contain some grazing are now quite useless. Some pastoralists try to find feed for their animals further south, where it is wetter and more vegetation grows. Many new waterholes have been dug, and **irrigation** schemes have been started in order to help these people. In most developments the nomads are expected to settle permanently, so that in some areas nomadism will gradually disappear.

nonrenewable resource A resource which cannot be replenished. **Fossil fuels**, for example, cannot be replaced or reused once they have been burned. The same is true of metals, although some scrap and waste metals can be collected and recycled. Metal ores are a finite resource which cannot be renewed.

WORLD NUCLEAR POWER STATIONS		
Percentage of electricity generated by nuclear power stations (top ten leading nations, 1995)		
	Country	Percentage
1	Lithuania	95%
2	France	77%
3	Belgium	56%
4	Slovak Republic	49%
5	Sweden	48%
6	Bulgaria	41%
7	Hungary	41%
8	Switzerland	39%
9	Slovenia	38%
10	South Korea	33%

North, the A region including the more advanced areas of the world – North America, Europe and Australasia – as defined by the **Brandt Commission**, which was set up in 1977 by Willy Brandt, the Chancellor of West Germany between 1969 and 1974.

North Atlantic Drift SEE **Gulf Stream**

North Atlantic Treaty Organization (NATO) A defence alliance formed in 1949 by Belgium, Canada, Denmark, France, Iceland, Italy, Luxembourg, Netherlands, Norway, Portugal, the UK and USA. Greece and Turkey joined in 1952, West Germany in 1965, Spain in 1982 and the Czech Republic, Hungary and Poland in 1999. Its headquarters are in Brussels, Belgium, and its original aim was to resist the perceived threat from Eastern Europe and the USSR. Although this threat has now all but disappeared, NATO still plays a major role and may send an army (made up of different forces from member countries) to monitor conflicts or troublespots in the world.

northing The eastwest grid line on a contour map. It is quoted second, after the **easting**, when giving a grid reference.

nuclear power station An electricity generating station which uses nuclear fuel as the source of energy. There are different types of nuclear power station, and although expensive to construct, they are quite cheap to run. They do not give off pollution like the thermal power stations, but present serious problems about the disposal of radioactive waste. Many have been built in isolated locations. In 1995, the USA ranked 19th in the world in terms of the amount of electricity generated by nuclear power stations (18%), with the UK was ranked 15th (27%).

nuclear winter If there were a number of nuclear explosions, large quantities of smoke and dust would be released into the **atmosphere**. These might intercept incoming solar radiation, thereby blocking the sun's light and heat and resulting in lower temperatures and wintry conditions at the Earth's surface.

nucleated settlement A small settlement which has grown up around a nucleus, such as a crossroads. Nucleated settlements are a cluster of dwellings; they are not spread out along roads in a linear fashion. Nucleated settlements often grow at a

spring or water hole, or on a patch of higher dry ground in a marshy area.

nuée ardente A cloud of hot gas which comes swiftly down the side of a **volcano** after an eruption. The cloud may also contain small dust particles and fragments of ash. It will kill anyone it passes over; in the Mt. Pelée eruption of 1902, all but one of the inhabitants of St. Pierre, on the island

NUCLEAR POWER STATION

A pressurized water reactor (PWR) is so named because the primary coolant (1) that passes through the reactor core (2) is pressurized to prevent it from boiling. The uranium-235 fuel is loaded into the reactor in pellets (3) contained by the fuel rods (4). To prevent an uncontrolled chain reaction, the fuel rods are separated by control rods of graphite (5). All the rods are loaded into the reactor from above (6). The primary coolant is heated by the fission reaction in the fuel rods and circulates into a steam generator (7), where it superheats the secondary coolant (8). The secondary coolant leaves the protective containment vessel (9) and drives turbines (10), which produce electricity through a generator (11). A third coolant loop (12) cools the secondary coolant, transferring the heat to a sea, river or lake. Reducing the temperature of the secondary coolant increases the efficiency of the transfer from the primary to the secondary coolant. The inset photograph shows the core (dark circular area) of a nuclear reactor during the "charging" period, at which time the first load of fuel is placed in the reactor core.

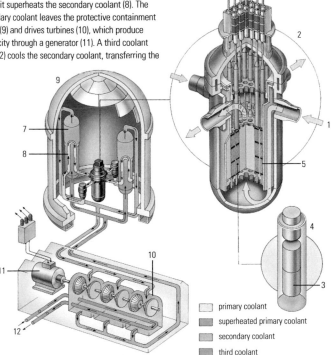

primary coolant
superheated primary coolant
secondary coolant
third coolant

of Martinique, were killed by the nuée ardente. There was also a nuée ardente at the eruption of Mt. St. Helens in Washington, USA, in 1980.

null hypothesis A statistical method used to determine the significance of the differences between two samples, or to frame a question which is likely to be proved incorrect by field studies or research. A null hypothesis is a negative assumption, when the researcher really believes that a positive answer is likely. Research may show that a null hypothesis is false, although this will not necessarily produce the final correct answer.

nunatak The **Inuit** word for a rocky outcrop projecting above the general level of an **ice sheet,** as in Greenland and Antarctica. It may be the top of a mountain peak.

Nunavut The new **Inuit** territory, created on April 1, 1999 in the former Northwest Territories of Canada. The territory makes up a fifth of the country, covering approximately 650 sq miles [2 million sq km]. It is the largest land claim ever settled in Canada.

nutrient cycle The transfer of nutrients in an **ecosystem** from one stage to another. Leaves fall to the floor to be broken down by bacteria and provide nutrients for the soil, which then sustains plant growth. The growing plants take energy from the sun, and provide food for any animals which live in the ecosystem.

O

oasis A fertile location which has water in an arid landscape. Usually ground water is brought to the surface in a well, but an oasis may occur at the point where a river flowing from a wetter region crosses the desert on its way to the sea, such as the Nile in North Africa or the Indus in southern Asia. Most Saharan oases contain large numbers of date palms, and there may also be many fields of cereals, vegetables and fruit. There is often a settlement at the oasis, sometimes with up to 30,000 inhabitants.

oats A cereal plant, *Avena sativa*, which can be used as a human food, in breakfast cereals and oatmeal, and also as an animal feed. The plant can survive in fairly wet environments, such as western Scotland, where rainfall totals may be over 40 in [1,000 mm], though the grain will not ripen in such locations. Oats can tolerate low temperatures in winter, and require summer temperatures of up to 60°F [15°C]. Oats are widely grown in Canada, northern USA, upland Britain, Norway, Sweden and other northern latitudes.

obsequent drainage A river or stream that flows in the opposite direction to the dip of the rock strata over which it is flowing.

obsidian SEE **volcanic rock**

occluded front A combination of a warm and a cold front in a **depression**. It occurs when the cold front catches up with the warm front and undercuts it. The effects of the warm front and the cold front together are to cause heavier and more prolonged rain than either front would give alone. **Cumulonimbus** clouds will be found at the occluded front. An occluded front is sometimes described as an "occlusion."

ocean The oceans cover nearly 71% of the Earth's surface. The Pacific Ocean is the largest, and it contains the deepest part of all the oceans. The others are the Atlantic, the Indian, the Arctic and the Southern Ocean, which surrounds Antarctica. The continental shelves are the shallowest parts of the oceans, and occur nearest to land. At the end of a **continental shelf** is a continental slope leading down to the deep sea plain, which covers vast areas between 10,000 ft to 20,000 ft [3,000 m and 6,000 m] deep. The ocean floor has many ridges and trenches; the Mariana Trench, in the Pacific, is the deepest at 36,161 ft [11,022 m]. SEE ALSO **deep**

ocean currents A distinctive and persistent movement of surface water. Most currents are the result of winds blowing the surface water. In the Atlantic Ocean there are equatorial currents, which flow from east to west because of the effects of the **trade winds**. The North Equatorial Current is just north of the Equator and it flows into the Gulf of Mexico. It turns to the right because of the shape of the land and the effects of the rotation of the Earth, and then flows across the Atlantic as the North Atlantic Drift. The current goes to Britain and Norway, but a mass of water turns right to head southward, past Portugal and along the African coast. At this point it is called the Canaries Current and is a cool current, not because the water is really cold, but because it is cool for its latitude. For the same reason, all currents which flow toward the Equator are cool currents. The Canaries Current rejoins the circulation to become part of the North Equatorial Current again. The other major current which joins this clockwise circulation is the cold Labrador Current; it heads southward along the west coast of Greenland, bringing icy water from the Arctic. It also brings icebergs, as far south as Newfoundland on occasions. In the South Atlantic there is a similar, though counterclockwise, circulation, and the cool current off the coast of southwest Africa is the **Benguela Current**. In the Pacific there are circulations which resemble those of

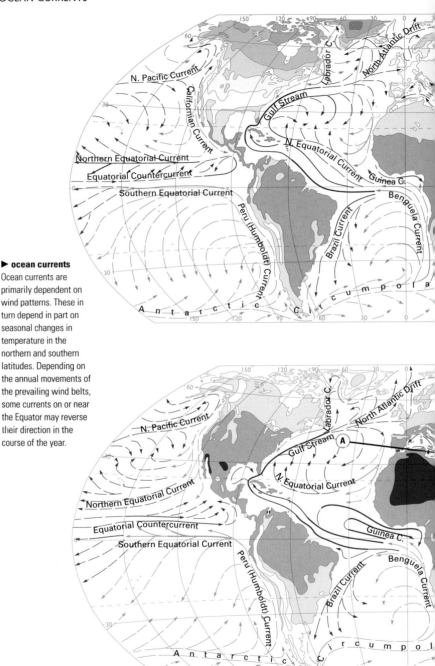

▶ ocean currents
Ocean currents are primarily dependent on wind patterns. These in turn depend in part on seasonal changes in temperature in the northern and southern latitudes. Depending on the annual movements of the prevailing wind belts, some currents on or near the Equator may reverse their direction in the course of the year.

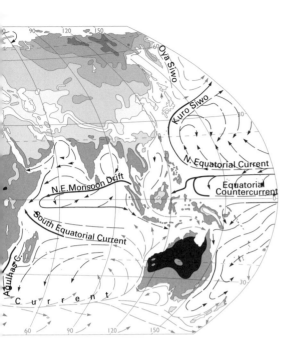

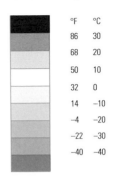

January
(northern hemisphere: winter)

actual surface temperature

	°F	°C
	86	30
	68	20
	50	10
	32	0
	14	−10
	−4	−20
	−22	−30
	−40	−40

ocean currents

cold	warm	speed (knots)
←– –	←– –	less than 0.5
←	←	0.5 – 1.0
←	←	over 1.0

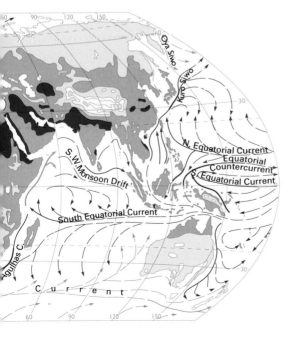

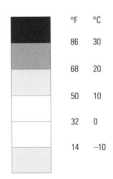

July
(northern hemisphere: summer)

actual surface temperature

	°F	°C
	86	30
	68	20
	50	10
	32	0
	14	−10

ocean currents

cold	warm	speed (knots)
←– –	←– –	less than 0.5
←	←	0.5 – 1.0
←	←	over 1.0

the Atlantic. The Indian Ocean is similar, but halfway through the year there is a reversal of flow, because the **monsoon** winds change direction. In the summer, while the southwesterlies are blowing, the current flows from west to east, but in winter the winds change to northeasterly, and the ocean currents flow from east to west. SEE ALSO **Gulf Stream**

ocean-floor spreading A movement of the seabed associated with **plate tectonics**. Along the **Mid-Atlantic Ridge**, for example, two plates are moving apart, and as a result a split is developing through the middle of Iceland. As the plates move, fissures open up and **basic lava** pours out; for this reason, Iceland is the most active volcanic area in the world at present. In the **Mesozoic** era, Western Europe and North America were close together, but have since been drifting apart for millions of years because of the movements of the ocean floor. Ocean-floor spreading is also taking place near Yemen and Somalia, and between the Nazca and Pacific plates off the west coast of South America. ALSO CALLED **sea-floor spreading**

ocean trench SEE **deep**

OECD SEE **Organization for Economic Cooperation and Development**

oil terminal A port where oil tankers carrying oil from producing regions unload in the importing country. The oil terminals have facilities for storing or refining the oil. Examples are Houston, Texas, USA, and Hardisty, Canada.

okta (*in meteorology*) A measurement of cloud cover equal to one eighth: eight oktas means that the sky is completely covered, and four oktas means that it is half covered.

O layer SEE **A horizon**

omnivore An animal which eats plants and animals.

onion weathering SEE **exfoliation**

oolite A type of limestone consisting of calcareous shells which accumulated on the seabed. Oolitic limestone was formed in the **Jurassic** period. ALSO CALLED **oolith**

oolith SEE **oolite**

OPEC SEE **Organization of Petroleum Exporting Countries**

open-pit mining Mining of rocks or minerals which are on or near the surface. There is often a layer of rock, called "overburden," on top of the mineral, which is removed and placed on one side, so that it can be replaced at the end of mining operations. Digging and excavating equipment are then used to remove the required mineral. Open-pit mining is generally much cheaper than **shaft mining** or **adit**. ALSO CALLED **open-cast mining**

open system A **system** into and out of which material and energy are passed. It is thus part of its environment, but regulates itself by homeostasis to maintain balance and a steady state. SEE ALSO **closed system**

optimum population The best population density for a particular region. It means there will be jobs for most people, and ample income and food supply. Some places have too many people for the available resources; for example, many parts of India, especially cities such as Calcutta, are overpopulated. Other regions may have fewer people than could be supported at a reasonable standard of living; for example, much of Australia. The optimum population of an area can change, if there are changes in sources of materials and income.

Ordovician The second-oldest period of the Paleozoic era, 500 to 430 million years ago. All animal life was restricted to the sea. Numerous invertebrates flourished, and remains of jawless fish are the first record of the vertebrates.

ore A type of mineral or rock which contains a metal or nonmetallic substance of some commercial value. Iron ore is likely to be a type of sandstone with a percentage of iron; if it is 25%, then of every 100 tons of rock dug up, 25 tons will be iron. Good-quality iron ore is 60% to 70% pure, whereas a good-quality copper ore may be no more than 1% pure.

organic Derived from the breakdown of vegetation or animal matter. The remnants of living matter form the organic part of soil, whereas the particles derived from the rocks form the inorganic content.

organic fertilizer Any fertilizer derived from organic material, such as manure, bonemeal, seaweed or fishmeal. It is thought to be better for the structure of the soil than chemical or inorganic fertilizers.

Organization for Economic Cooperation and Development (OECD)
An organization that took over in 1961 from the Organization for European Economic Cooperation (OEEC) and has 29 member countries (including Australia, France, Italy, Spain, the UK and USA). Its aims are to promote mutual growth and development, develop common social welfare policies, and help developing countries.

▼ **Ordovician** The most significant movement of the plates during the Ordovician period was the movement of North America towards n Europe, thus compressing the sea area between the two land masses.

Organization of Petroleum Exporting Countries (OPEC)
An organization set up in 1960 which, as a central body, fixes the price of oil on the international market at regular intervals. Its founder members were the chief oil-producing countries – Iran, Iraq, Kuwait, Saudi Arabia and Venezuela – which hoped to take back control over the oil industry; twice, in 1973 and 1974, they increased oil prices so dramatically that the economies Western countries suffered greatly.

orogenesis The process of mountain formation. As a result of plate movement, folding and faulting are likely to occur as the crustal materials are compressed and forced up. The major periods of mountain formation have been the Caledonian, Hercynian and Alpine. The **Caledonian folding** occurred at the end of the Silurian period, and good examples can be seen in Scotland and Norway. The **Hercynian** folding occurred at the end of the **Carboniferous** period, forming the Harz Mountains in Germany, the Massif Central in France and the Meseta in Spain. The Caledonian and Hercynian are referred to as "old" fold mountains. The **Alpine folding** took place during the **Tertiary** period; mountains formed at this time are referred to as "young" fold mountains, such as the Rockies of North America. ALSO CALLED **orogeny**. SEE ALSO **fold mountains**

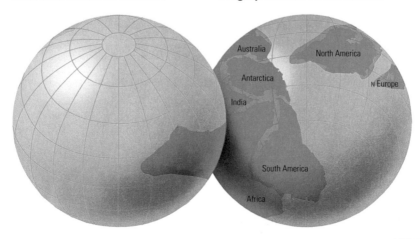

orogeny SEE **orogenesis**

orographic rainfall Rainfall which occurs as a result of winds rising to pass over high ground. As the winds rise, the air is cooled, causing condensation and clouds begin to form. If the air continues to rise, rain will fall. In temperate latitudes the total rainfall will increase with height, up to about 6,500 ft [2,000 m]. Above that height, the air becomes so cold that it cannot hold as much water vapor, and so the rainfall quantities begin to decrease. The winds which cool as they ascend mountains warm up as they descend on the leeward side, and are therefore able to hold more moisture. So, the rainfall totals are much lower on the leeward side of the hills, and it is referred to as a "rain shadow." All mountains are likely to have an area of rain shadow. In England it is on the east side, but in Australia it is on the west; for example, while the area of Queensland to the east of the Great Dividing Range receives 30 in to 100 in [800 mm to 2,500 mm] annual rainfall, the west side totals 10 in to 20 in [250 mm to 500 mm] per annum. ALSO CALLED **relief rainfall**

orthoclase SEE **feldspar**

outcrop A part of a layer of rock or rock formation which is exposed at the surface of the ground.

outlier A mass of newer rocks surrounded by older rocks, often because it has been detached from a larger formation by **erosion**.

out-of-town location A site in a rural area near to a city, in which new development can take place. Several new industries have been created in out-of-town locations, where there is more space for building, land may be cheaper, and there may be less traffic congestion. Many new malls and supermarkets have been built in out-of-town locations, where there is plenty of space for car parking.

outport 1 An isolated fishing village on the coast of Newfoundland, Canada. 2 British name for a secondary port which is downstream, nearer to the sea than the main port. Sometimes the main port has

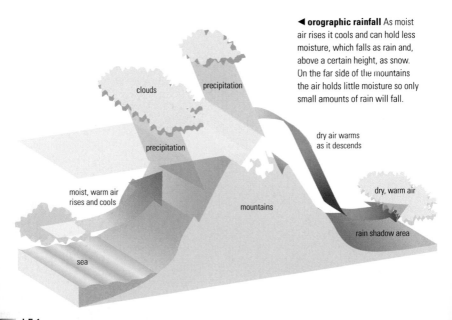

◄ **orographic rainfall** As moist air rises it cools and can hold less moisture, which falls as rain and, above a certain height, as snow. On the far side of the mountains the air holds little moisture so only small amounts of rain will fall.

precipitation

clouds

precipitation

dry air warms as it descends

moist, warm air rises and cools

mountains

dry, warm air

rain shadow area

sea

declined because modern ships are unable to get as far upstream as the smaller boats of the past.

output The end product of a process or system; for example, the crops or animals produced on a farm. SEE ALSO **input**

outwash plain A depositional area built up by the sands and gravel brought down by streams flowing from a glacier or an ice sheet. The materials have literally been washed out by **fluvioglacial** activity. They are often sorted, and so the largest particles are deposited first, the materials becoming finer and finer moving downstream. Broad plains, called "sandur" in Iceland, are sometimes built up by several streams flowing from an icecap. There are outwash deposits on the "Corn Belt" of the USA. They are often sandy and infertile.

overburden SEE **open-pit mining**

overcropping When too many crops have been planted continually, the soil may lose its fertility, since it has had no chance to restore it naturally between crops.

overcrowding Too many people or organisms in too little space. It is particularly applied to large numbers of people living in one dwelling.

overflow channel A valley formed by water overflowing from a lake. Examples are common where lakes are found in hills or mountains.

overfold A fold in which the rock strata have been pushed right over because of intense compression.

overgrazing The excessive use of an area of land for pasture. If too many animals are allowed to graze an area, they eventually remove all the grass and leave the soil exposed. Once this happens the soil is likely to be eroded, and no more grass will grow. Overgrazing can destroy the land and turn it into desert. This has been happening in the **Sahel** in central Africa, with the result

that the Sahara Desert is spreading southward. Cattle and sheep can cause overgrazing, but goats are even more notorious for removing all the vegetation and leaving the soil bare. To bring about a reduction in overgrazing is difficult, because animals are the only source of wealth, food and income for many people. More feed could be grown for the animals if irrigation schemes were set up, and this would enable some of the pastures to be rested and restored. Rotation of pastures would be vital in areas where overgrazing has taken place.

overpopulation The condition of having too great a population for the available

OXBOW LAKE

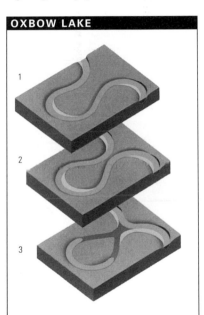

Small variations in the speed of water flow can alter the course of a river. Water flowing at the outside of a bend (1) moves more quickly and cuts away the bank as it picks up sediment. On the inside of the bend the water moves more slowly and deposits sediment. Over time a meander (2) forms. The river soon cuts an alternative course across the neck of the meander (3). Sediment builds up at the ends of the loop, which eventually becomes separated, forming an oxbow lake.

resources. If a country or a region is unable to maintain a reasonable standard of living for its inhabitants, it can be described as overpopulated. It is not necessarily related to the density; some parts of the USA, for example, have high densities of population, but since everyone is relatively wealthy and well fed, the population is not too high. On the other hand, parts of the **Sahel** in central Africa are quite sparsely populated, but the land is too poor to support the population. The situation could be reversed if, for example, better farming techniques were introduced. Equally, there could be a reversal in wealthy countries. If all the minerals ran out and industry declined, then they could reach a state of overpopulation. SEE ALSO **optimum population**

overspill A proportion of the population of a city which moves out to find more space or better living conditions in a new area a few miles away. Sometimes a new town or suburb is created to cater for the overspill or an existing town may be enlarged.

overthrust SEE **folding**

oxbow lake A small lake which was originally a **meander** of a river. As sediment is deposited, the meander becomes cut off from the river to create a lake. Once formed, the oxbow lake will gradually shrink as sediment fills it in; vegetation will grow on the new muddy area, and the land can be reclaimed. ALSO CALLED **cut-off, horseshoe lake**

ozone A form of oxygen which is found in very small quantities in the **atmosphere**, mostly in the ozone layer, about 20 miles to 35 miles [30 km to 60 km] above

OZONE

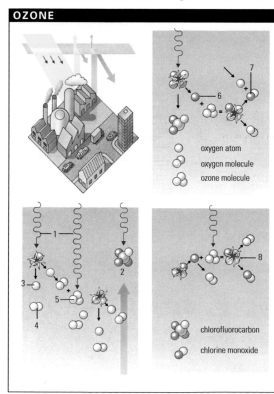

oxygen atom
oxygen molecule
ozone molecule

chlorofluorocarbon
chlorine monoxide

A naturally occurring substance, ozone acts as a sunscreen for the Earth because its molecules absorb the Sun's ultraviolet radiation (1). The presence of chlorofluorocarbon pollution (2) causes the ozone layer to break down allowing ultraviolet rays through. Ozone is created when ultraviolet rays split oxygen molecules. The lone oxygen atoms (3) bond with oxygen molecules (4) to make ozone (5). When, however, chlorofluorocarbons are present, they are split by the ultraviolet rays. The released chlorine atom (6) in turn splits ozone molecules to form a chlorine monoxide molecule (7) and oxygen. The process is continued as the chlorine monoxide absorbs the lone oxygen atoms that previously formed ozone. This frees the chlorine atom, which splits another ozone molecule (8) creating another chlorine monoxide molecule and oxygen.

the Earth's surface. The ozone layer absorbs much of the ultraviolet wave energy, but allows longer ultraviolet waves to pass through. Gaps have been seen in the ozone layer in recent years (thought to be caused by man-made products, such as the pro-pellant gases in aerosol sprays), and this could cause serious changes in climatic conditions. The additional ultraviolet waves which could get through to Earth might also constitute a health hazard, increasing the risk of skin cancer.

P

Pacific coast SEE **concordant coast**

paddy I The rice plant, *Oryza sativa. 2* A wet field in which rice is grown. Not all rice is grown in paddy fields, since some varieties can be grown in quite dry areas. SEE ALSO **rice**

pahoehoe The Hawaiian name for **ropy lava**. SEE ALSO **aa**

Paleocene A geological epoch that extended from about 65 to 53 million years ago. It is the first epoch of the Tertiary period, when the majority of the dinosaurs had disappeared and the small early mammals were flourishing.

paleoclimatology The study of prehistoric climates. Records of past climates are found in sedimentary rocks, in cores taken through deep layers of ice, and in fossil-bearing cores from the beds of seas and lakes. From such evidence, climatologists

have discovered that the Earth is subject to alternate periods of cold, called glacials or **ice ages**, and warmth, called interglacials. During the last two million years there have been 17 ice ages.

paleomagnetism The study of old magnetic fields and the magnetism which can be found in ancient rocks. When **igneous** rocks cool and harden, they retain a certain magnetization, resulting from the presence of particles of iron oxide which orientate themselves parallel to the Earth's magnetic field at that period. This enables geologists to work out the location of these rocks at the time of their formation and can be very helpful in studies of drifting continents and **plate tectonics**.

paleontology The study of fossils. It is more relevant to geology than geography, but fossils are seen in studies of chalk and limestone rocks, and also in coal measures, where there are many plant fossils. The study of fossils, and the plants and animals which created them, is an important way of finding out what conditions were like on the Earth millions of years ago, since the requirements for growth of the different plants and animals are known. For

▼ **Paleocene** The drifting apart of continents continued during the Paleocene epoch. The movement between Africa and Europe raised the Alps. A great area of volcanic activity reached from northwest Europe towards the position of Iceland.

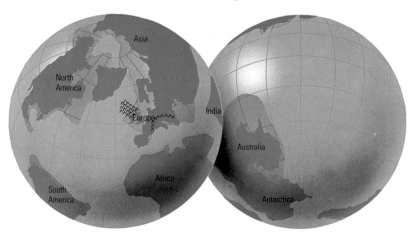

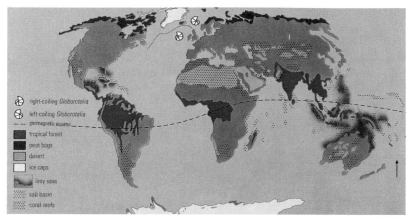

right-coiling *Globorotalia*
left-coiling *Globorotalia*
– – – geomagnetic equator
tropical forest
peat bogs
desert
ice caps
limy seas
salt basin
coral reefs

▲ **Paleoclimatology** Much can be revealed about ancient climates through the study of traces left in contemporary rocks. The major present-day formations of such preservable climate-related features are shown. The foraminiferan *Globorotalia* is an indicator of sea temperature. It coils to the right in warm waters and to the left in cold waters. Coral reefs and major carbonate deposits are both typical of warm, shallow seas. Common to desert environments are evaporite deposits (salt basins) and reddish-hued sandstones. The lush plant life of tropical forests and swamps is the raw material from which coal is formed. Ice sheets groove and scratch the face of rocks and leave characteristic deposits of glacial till, and peat bogs are typical of the tundra environment along the fringes of the icecaps.

example, fossils have been found on Mt. Everest which reveal that some of the rocks forming the world's highest mountain were in a warmish sea about 40 or 50 million years ago.

Paleozoic The geological era which lasted from about 570 million years ago until about 225 million years ago. It is made up of six periods: **Cambrian**, Ordovician, Silurian, Devonian, Carboniferous and Permian. During the Cambrian, Ordovician and Silurian periods, many current land areas were covered by sea, and trilobites, graptolites and brachiopods were the commonest animals. New mountains were formed at the beginning of the Devonian period, and then coal deposits were laid down in the Carboniferous period (a combined named for the Mississippian and Pennsylvanian periods). This was followed by the Permian period when the Appalachians were thrust up. The word "Paleozoic" means "ancient life," and the era contained fairly primitive animals only. It was followed by the Mesozoic era. ALSO CALLED

primary era SEE ALSO **geological column, geological time scale**

pampas An area of temperate grassland near the River Plate estuary, lying mostly in Argentina and also in Uruguay. The name is derived from the Spanish word for plain. With temperatures of around 70°F to 75°F [20°C to 25°C] in January and 50°F to 60°F [10°C to 15°C] in July, and an annual rainfall of 20 in to 40 in [500 mm to 1,000 mm], conditions are ideal for grass. The early settlers reared sheep and cattle on the rich pastures, and there were local cowboys, called *gauchos*, to look after the herds of animals. Nowadays, much of the grassland has been plowed, and there are vast fields of corn, flax, wheat and alfalfa. Large cities such as Buenos Aires and Montevideo have grown up in this region.

Pangea The name given to a vast land mass which millions of years ago was the only land on the Earth. It consisted of the **sial,** which gradually split to form the ancient continents of Gondwanaland and

Laurasia. Pangea began to break up about 200 million years ago; it was first written about in 1912 by Alfred Wegener in connection with his theory of **continental drift**.

parallel drainage A **drainage** pattern where the channels of streams and their tributaries run almost parallel to each other.

parasite An organism which lives on or in another living organism and draws most of its food, and therefore energy, from it, sometimes causing harm to the host organism. For example, a mosquito, which lives on the blood of animals and humans.

parent material The **bedrock** from which a soil has been formed; it will be weathered, but little altered.

passive glacier A glacier with low rates of accumulation ("alimentation") and depletion (**ablation**) because it receives little snow to feed it in the winter and little melting in the summer. It therefore flows very slowly, and erosive and transportation processes are minimized.

pastoral farming The rearing of herbivorous animals. In addition to grazing on pasture, many animals have to be fed, especially in the winter months. There are a number of different types of pastoral farming, including the rearing of beef cattle on the plains of Texas, USA; sheep farming in the Welsh hills, the Pennines in England, or the Murray-Darling basin of Australia; dairy farming, as in Denmark and New Zealand. Some pastoral farmers in parts of Africa, such as the Masai in East Africa or **Fulani** in West Africa, are nomadic and wander in search of grazing for their animals. In other regions around the world, there are pastoralists who rear reindeer, llamas, yaks or goats.

pasture An area of grassland suitable for grazing by animals, such as sheep, cattle, goats, llamas, yaks, alpacas or deer. Most important and most numerous are sheep and cattle. Sheep can survive in areas of poor pastures, such as hilly or semiarid areas. They do not produce good meat in such environments, but they yield fine-quality wool. Sheep also thrive in richer pastures and wetter areas, where good meat as well as wool can be produced. Dairy cattle are found only in areas of rich pasture, as they need to be well fed at all times, but beef cattle can live in poorer areas. Pastures in temperate humid landscapes are generally very nutritious for animals. Other areas, such as the **pampas**, are also very suitable, but drier areas, such as the high plains of Kansas, Wyoming and elsewhere in the western USA, can have very poor pasture. In tropical lands, the grassland is often quite rich in summer, but very dry, brown and shriveled in the winter months.

pays Low plateaus of land in northern France with their own distinctive characteristics. For example, the Beauce, which has fertile soils, and the Sologne, which has poor, sandy soils. Each of these regions will then develop its own culture based on its particular features.

peasant farming Small-scale farming, in which farmers grows crops and rears animals mainly for their own use, but probably with a small surplus for sale. It is intermediate between **commercial agriculture** and **subsistence farming**, but closer to the latter. Some peasants are very poor, as in parts of West and East Africa, but others may be leading quite a comfortable life; for example, in many parts of France. A peasant farmer usually has very close links with the land, and looks after it very carefully.

peat A soil composed of dead and decaying remains of vegetation. It is generally dark brown or black in color, and is thought to be similar to the first stage in the formation of **coal**. It forms in areas of high rainfall or very poor drainage, and contains a high proportion of water. In very wet conditions, plants do not decay when they die, because the conditions are nearly **anaerobic**, and so the vegetation is not broken down by bacteria. Waterlogged and airless conditions create a very acidic and infertile landscape, which can become very boggy. Only specialized plants, notably sphaghnum moss, can survive in such areas. Large tracts of upland Britain are covered by **blanket bog**, which is a large sheetlike expanse. The low-lying peaty areas, such as in the Fens, can be drained more easily, and the black soils can be quite fertile and productive. This is because the plants which formed the lowland peat are less acidic and contain more nutrients. SEE ALSO **bog**

pedalfer A large group of soils which contain aluminum ("al") and iron ("fer"). The zonal classification of the world's soils contains two groups: the pedalfers and the pedocals. Pedalfers are found in the more humid regions, such as in many parts of the tropics, where **laterites** are a common type of pedalfer. Red and yellow soils are a variety in the southeastern USA. In maritime regions, such as Britain, brown forest soils are common, and in the cooler coniferous forests of Canada, Norway, Sweden, Finland and northern Russia, there are widespread **podzols**, another common type of pedalfer. SEE ALSO **pedocal**

pedestal rock A rock which is shaped like a mushroom as a result of **corrasion** by sand. In arid landscapes, windborne sand is moved along just above ground level. The sand erodes any rocks which are standing above the surface of the land, and as the corrasion is most active 3 ft [1 meter] or so above ground level, the eroded rocks become mushroom- or pedestal-shaped. Pedestal rocks occur in most deserts; examples can also be seen near some coasts and on hills. ALSO CALLED **mushroom rock**. SEE ALSO **zeugen**

pediment An area of eroded fragments of rock close to uplands in a semiarid or arid region. The accumulated fragments gradually build up a plain which slopes down gently from the mountains. The largest rocks and boulders are generally found nearest to the mountains, and the particles become progressively smaller as

the distance increases away from the high ground. SEE ALSO **pediplain**

pediplain A large expanse in a desert or semiarid region which is fairly low-lying and flat. It is formed by the merging of several areas of **pediment**.

pedocal Any of a large group of soils which contain a high proportion of calcium. Pedocals are found in dry regions, such as the drier parts of the **steppes** and **prairies**, and the semiarid lands of the world. They are one of the two main groups of zonal soils, the other group being the **pedalfers**. Pedocals do not suffer much from **leaching**, because they only occur in areas where the annual precipitation is about 20 in [500 mm] or less.

pedology The study of soil.

pelagic Occurring in, or inhabiting, the open sea. Certain fish are described as "pelagic fish" because they live in the moderately deep parts of the ocean as opposed to the shallow areas of **continental shelf**. Pelagic deposits are fine muds which accumulate on the bed of the oceans. They consist mainly of the dead remains of tiny sea-dwelling plants and animals. Remains of this kind will also fall to the seabed in shallower waters, but are lost among the greater deposits of land sediments brought to the sea by rivers, flowing from adjacent land masses. The fine muds are called "oozes," some of which are predominantly calcareous while others are mainly siliceous. Pteropod and globigerina oozes are the commonest calcareous muds, while radiolarian and diatom oozes are the commonest siliceous muds. There are also fine muds called "red clays," which are formed from dust deposited from the **atmosphere** after a volcanic eruption.

Peléan Relating to Mt. Pelée, an explosive and very acidic volcano on the island of Martinique in the Caribbean, which had a violent eruption in 1902. It sent a cloud of gas (**nuée ardente**) down the mountainside after the explosion. The gas killed everyone in the nearby town of St. Pierre, with the exception of one man, who was in a prison cell beneath ground level, awaiting execution. He was later rescued and reprieved. SEE ALSO **volcano**

peneplain An old-age landscape, such as the Canadian Shield in North America. When mountains and plateaus have been eroded for millions of years, they are worn down until they are low and fairly flat. The more resistant hills may be left standing up above the general level of the plain, as **residual hills** or **monadnocks**. The peneplain will have been reduced almost to a plain by the effects of all the agents of **denudation**, but especially by the action of rivers. A peneplain is the final stage in the **cycle of erosion**, which wears large mountain ranges down until they end up as low landscapes.

peninsula A narrow strip of land which protrudes into the sea. It will be attacked by marine erosion on both sides, and **joints** are likely to be enlarged to form **caves**. Peninsulas often have cliffs along one or both sides.

per capita income The income per person, generally quoted over a period of one year. The income of a country, or region, is divided by the number of people living in that country, to give an average income for every wage earner. The mean, or average, per capita income for a year is often used as an indication of wealth and economic development in a country. In many developed countries, the per capita income is more than $20,000 dollars per annum, whereas in many African countries, for example, it is well below $1,000 per annum.

percolation The process by which water passes down through the pores in the soil and through cracks and joints in the rocks. The water moves downward under the influence of gravity. The rate of percolation depends on the rock type, as well as on the amount of rainfall. Percolating water can carry dissolved chemicals, and may cause **leaching**.

percoline A line of water seepage through soil; part of the general **throughflow** movement. Percolines follow the general slope of the surface, and may have formed along lines created by worms or burrowing animals. There may be several percolines, which are roughly parallel, or they may have formed in a more dendritic pattern. They are situated just beneath the surface.

perennial irrigation SEE **irrigation**

perennial stream A stream or a river which flows permanently. SEE ALSO **exotic stream**, **intermittent stream**

periglacial Around the edge or on the periphery of a glacial area. A periglacial landscape was not covered by ice, but was affected by cold conditions such as those now experienced in the tundra areas of northern Canada or northern Russia. During the periglacial period there was **permafrost**, leading to the occurrence of **frost heaving**, **freeze-thaw** activity and **solifluxion**. Other periglacial features include stone circles and stone stripes, which can be seen in the **tundra** areas of the Arctic. Also, there are accumulations of combe rock, which are the result of solifluxion. SEE ALSO **combe**

periphery The edge or outside. The core-periphery idea is based on the studies of economic development, which suggest that the main cities, especially capitals, are the core regions, where most money, industry and wealth can be found. The more isolated or peripheral areas tend to receive less investment and are not as developed. SEE ALSO **core** [def. 1]

permafrost Land that is permanently frozen, often to a considerable depth. The top few inches will generally thaw out in the summer, but the **meltwater** will not be able to sink into the ground because of the frozen **subsoil**. Drainage takes place only if the water can flow down a **gradient**. If the landscape is fairly flat, there will be surface water lying on the ground throughout the summer. This is found in many **tundra** areas, and the watery surfaces are excellent breeding grounds for mosquitoes and other insects. The insects attract many birds, which spend the summer in these areas, but migrate south for the winter. The wet surface layer causes serious problems to economic development, because the water freezes and thaws repeatedly as the temperatures change from day to night, causing the surface of the land to move and heave. Road building, pipe-laying, house construction, etc., become very difficult, and many different methods have been employed in Russia, Canada and Alaska, USA, to overcome the problems.

permanent pasture An area of grassland used for feeding cattle or sheep for many years.

permeable A term denoting rocks which are **porous** and absorb water. Water can percolate through the pores down to the **water table** to provide a supply of **ground water**. Most rocks are permeable.

Permian A geological period of the Paleozoic era lasting from 280 to 225 million years ago. There was a widespread geological uplift and mostly cool, dry climates with periods of glaciation in the southern continents. Many groups of marine invertebrate animals became extinct during the period.

personal space That area around a person that he or she tries to keep for himself or herself to allow conversation and normal movement to take place without feeling overcrowded. The amount of personal space a person requires varies between different cultures.

pervious A term denoting rocks through which water can pass by moving through cracks and fissures. **Carboniferous** limestone and **granite** are good examples of pervious rocks.

pesticide A chemical substance used to kill insects or animals that are regarded as harmful to human beings and their activities; for example, if they are destroying crops.

petrochemical industry A chemical industry which uses petroleum as its main raw material. A wide range of products can be made from petroleum, including many plastics. Petrochemical industries are often located near oil refineries, for example, Houston, Texas, in the USA.

▼ **Permian** During the Permian period northern Europe collided with southern Europe, pushing up the Variscan-type fold mountains. This combined block began to move towards the Siberian plate.

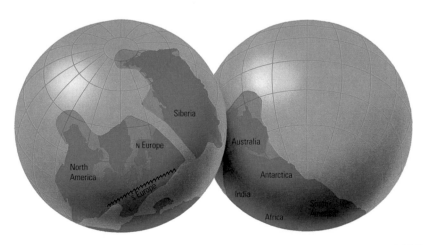

Siberia

N Europe

North America

S Europe

Australia

Antarctica

India

South America

Africa

PESTICIDE

Applied to the land, pesticides, even in small doses, are poisonous to many animals. The concentration of some pesticide poisons increases along the food chain, finally becoming lethal to animals at the end of the chain. A pesticide (1), such as DDT, is applied to water at 0.015 parts per million (ppm) to control midge larvae, but the plankton (2) accumulates it at 5ppm. The fish population (3,4) builds up still higher concentrations, and finally a grebe (5), which feeds on the fish, accumulates as much as 1600ppm of the pesticide in its body fat – enough to kill the bird.

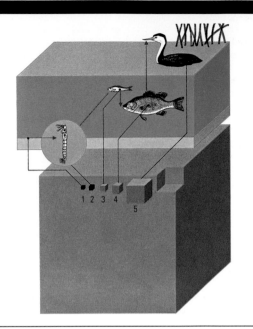

pH A measure of soil acidity, in which 7.0 is neutral, 6.0 or less is regarded as **acidic** and 8.0 is quite **alkaline**. It is a simple method of classifying soils, and is increasingly used by keen gardeners. **Cereals** grow best where the pH is about 6.5.

photosynthesis The chemical process where green plants build up complex organic compounds from atmospheric carbon dioxide and water using sunlight as energy (absorbed by the plants using chlorophyll). Oxygen is released as a result. Since all plants and animals are directly or indirectly supplied with the energy they need to live through the products of photosynthesis, it is a vital process.

phototroph An organism that obtains energy from sunlight.

physical geography One of the two major parts of geography, the other being **human geography**. Human and physical studies are often closely linked because of the effects that physical features can have on man, and the way that man can affect the landscape. Physical geography includes such subjects as **biogeography**, **ecology**, **geomorphology**, **hydrology**, **meteorology**, and **pedology**.

physical weathering SEE **weathering**

physiography SEE **geomorphology**

pie chart A circular diagram divided into segments which represent different percentages of a set of information. ALSO CALLED **pie graph**

piedmont Relating to the foot of a mountain. The plateau at the eastern margin of the Appalachian Mountains in the USA is called the Piedmont Plateau, and there is also a Piedmont (Piemonte) region in northern Italy. When two or more glaciers emerge from valleys on to a plain, and spread out until they merge with their neighbors, the resulting icefield is called a "piedmont glacier," examples of which can be seen in Alaska, USA.

pig iron The halfway stage between iron ore and the finished product, steel.

pike A summit of a hill in the Lake District and other locations in northern England. It is generally a rocky and fairly rugged summit.

pillow lava Lava that has occurred when magma is expelled under water, or flows into water before solidifying. The water therefore cools the outer layer quickly and the lava formed is rounded, making it look like a mass of jumbled pillows.

pingo A cone-shaped mound that stands alone and has a core of ice. It is formed when freezing water expands and is forced upward. This may happen in areas of **per-**

▼ **pillow lava** When a lava flow enters the sea it breaks up into a number of globular structures. These structures may come to rest piled on top of one another. The same formation occurs if magma is expelled directly into the sea.

mafrost where water cannot reach the surface and therefore pushes up the ground as it freezes.

pipeline A continuous length of piping for transporting liquids or gases. Pipelines are used to carry oil from many oilfields to ports or to refineries – for example, the Alaska pipeline in the USA – and some natural gas is transported by pipeline. Water is carried by pipes from reservoirs or rivers to houses and factories, and in irrigated areas it may be piped to the fields in order to reduce the loss from evaporation, which is enormous if open canals are used.

placer A deposit of sand and gravel containing particles of a precious mineral, such as gold, tin or platinum. Further upstream, river erosion will have created sediment which is then transported down the valley. The accumulated sands often contain particles of heavy minerals, which can be obtained by panning. In the late 19th century, there were gold rushes to California and Alaska, USA, the Yukon Territory in Canada, and to many other places in search of deposits of this kind,

metal-bearing rock
weathered by rain

debris washed
down to the shore

river and ocean currents ensure heavy metal
particles accumulate and form deposits

▲ **placer** Deposits occur in rivers, but much placer is washed out to sea. Most offshore mineral deposits, with the exception of oil and gas, do not occur in sufficiently high-ore grades to warrant their economic production. Nevertheless significant amounts of tin, diamonds, gold and titanium are recovered from beach and offshore placer deposits all over the world. Sulfur is extracted commercially in a number of areas, including the Gulf of Mexico. The most important mineral resources after oil and gas are sand and gravel dredged from offshore deposits for use in the construction industry. Placer deposits occur where metal-bearing rock on land is weathered and the debris produced is washed down to the sea by rivers. There it is sorted by the currents, waves and tides so that the heavy metal particles accumulate to form deposits of mineral sand. These typically take the form of beach deposits, but where the sea level has changed they can be found well out on the continental shelf. The sands are lifted by dredgers and sifted for their metal content.

and tin is still obtained from placer deposits in Malaysia.

plagioclase SEE **feldspar**

plain An extensive area of low-lying, flat or gently undulating land. A prime example would be the Great Plains in the west central USA. Some plains are the result of **denudation** wearing away higher land to form a **peneplain**, such as the Canadian Shield in North America. Most plains, however, are the result of **deposition**. Some of the flattest areas on Earth are in regions where land has been reclaimed from the sea, as in the Fens in England, and the **polders** in the Netherlands. Other very flat areas are **lacustrine** plains, where former lakes have been filled in with sediment. An example is the Red River valley to the south of Lake Winnipeg, in southern Canada and northern USA. Flood plains or alluvial plains are in river valleys, and glacial plains occur where large expanses of **till**, or glacial drift, have been deposited to cover the landscape, as in Norfolk in England, and eastern Denmark. Plains can be covered by any type of vegetation; for example, in South America, the Amazon lowlands are formed of an alluvial plain covered by forests. There are several large plains which are covered by grasslands, and their names are sometimes used both for the type of vegetation as well as the geomorphological landscape feature; for example, **prairies, steppes** and **pampas**.

planetary winds The major winds of the world, which blow over very large areas. In tropical latitudes these are the **trade winds**, which are very persistent. In the northern hemisphere, the trades are north-easterly, and blow from areas such as the Sahara toward the Equator. In the southern hemisphere they are south-easterly. Both the north-easterly and south-easterly trades blow from the tropical high-pressure zones (the **horse latitudes**) toward the equatorial low-pressure zone (the **doldrums**). On their journey they are affected by the rotation of the Earth, which deflects northern hemisphere winds to their right, and southern hemisphere winds to their left. In temperate latitudes, the planetary winds are the westerlies, which are far more variable than the trade winds. In the northern hemisphere they are south-westerlies, and in the southern hemisphere they are north-westerlies, known as the "Roaring Forties."

plankton The very small plants and animals which float or drift around in the sea. Many of them are so small as to be invisible to the naked eye, and many are young plants or animals which will grow much bigger. They are the main source of food of many larger creatures, and are most abundant in the shallow areas of continental shelves, where sunlight can penetrate to provide food and energy. Plankton are often especially numerous in areas where cool and warm water meet. Such areas include the Grand Banks off Newfoundland, the North Sea around Britain, and near the Humboldt Current off Peru, which have been among the most prolific fishing areas.

planned economy An economy that is controlled by the central government; i.e. capital, labor, production and distribution of goods are planned, usually to follow a set economic plan. Planned economies are often found under communist governments and are the opposite of free-market economies.

plantation A large farm or estate, especially in the tropical and semitropical latitudes, where the commercial production of one crop takes place. Many plantations were set up and financed by foreign capital, with foreign companies organizing the production of such commodities as rubber, cocoa, bananas and sugar cane. In the past, many plantations used slave labor, notably the cotton plantations of the southern USA, but plantations today often provide a good source of employment for local people. Most of the plantation crops are for export, especially to North America and Western Europe, and, as much of the plantation farming is organized by large international companies, many of the profits also go to North America and Western Europe. Plantations often take up good farm land, which could be used for growing food crops for the local people. Positive features of the plantations are that, in addition to providing employment, they have often provided schools and hospitals, paid for housing for the workers, and probably contributed to the construction of roads in the area. Other aspects of **infrastructure**, such as water supply or electricity, may also be developed by the plantation company. Major plantation areas include coffee on the Brazilian plateau, rubber in Malaysia, bananas in the West Indies and Central America, cocoa in Ghana and Nigeria, and tea in India, Sri Lanka and China.

plateau An elevated area of land which is fairly flat. The plateau may be broken by deep river valleys, making it a "dissected plateau," such as the Meseta of Spain. Mountain ranges often stand up above the general level of the plateau, as on the Meseta or the plateau of Brazil. Some plateaus may be very high, over 10,000 ft [3,000 m] above sea level, as in Tibet, China. Even though they are high, plateaus can be surrounded by mountains, and are then called **intermontane** plateaus; there are examples in Peru, Ecuador and Bolivia in the Andes. Some plateaus may be the result of uplift between faults to form block mountains, such as the Massif Central or the Vosges in France. The plateaus of the southwest USA consist of almost horizontal strata; this has contributed to the formation of large **canyons**, which dissect the plateaus.

plate tectonics The study of the formation of the major structures on the Earth's surface by the movement of the underlying plates. If they move together, the colliding plates may cause earthquakes, volcanoes and folding, but if they move apart, basaltic lava eruptions are the most likely result. The old theory of drifting continents is now related to the movement of the plates, some of which carry continents as though they were passengers on a raft. At "constructive margins," where plates are diverging, i.e. moving apart, as along the **Mid-Atlantic Ridge**, especially in Ice-

PLATE TECTONICS

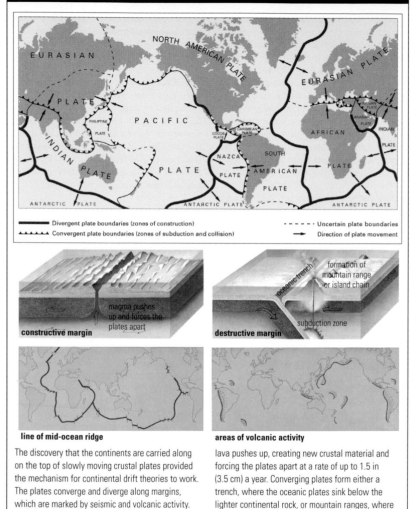

Divergent plate boundaries (zones of construction)   - - - - - Uncertain plate boundaries
▲▲▲▲▲ Convergent plate boundaries (zones of subduction and collision)   ⟶ Direction of plate movement

constructive margin — magma pushes up and forces the plates apart

destructive margin — oceanic trench, formation of mountain range or island chain, subduction zone

line of mid-ocean ridge

areas of volcanic activity

The discovery that the continents are carried along on the top of slowly moving crustal plates provided the mechanism for continental drift theories to work. The plates converge and diverge along margins, which are marked by seismic and volcanic activity. Plates diverge from mid-ocean ridges, where molten

lava pushes up, creating new crustal material and forcing the plates apart at a rate of up to 1.5 in (3.5 cm) a year. Converging plates form either a trench, where the oceanic plates sink below the lighter continental rock, or mountain ranges, where two continents collide.

land, there are frequent extrusions of **magma** to form new basaltic rocks. Seafloor spreading is occurring in these locations. At convergent boundaries ("destructive margins"), one plate sinks below the other in the **subduction zone**; this process is happening along the western edge of the Pacific Ocean. The plates are affected by

the convection currents in the Earth's mantle, and really belong to the field of geology. However, because of their influence on mountain formation, they are also a very important part of geography.

Pleistocene A geological epoch, which takes its name from Greek words meaning "most recent." It can be used to refer to the last 2 million years including, or excluding, the most recent time since the last ice advance. The postglacial period is called the **Holocene**. The beginning of the Pleistocene is now generally agreed to be about 2 million years ago; in some places it coincides with the beginning of the Ice Age, but as the ice advanced over different places at different times, there is no set time for the beginning of the glacial phase. In North America, the horse and camel became extinct during the Pleistocene epoch. SEE ALSO **geological column, geological time scale**

plucking A process associated with glacial conditions. When ice in a glacier freezes to a rock on the valley floor or side, the movement of the ice is likely to pull part of the rock away. Plucking gives rise to jagged rock surfaces. It is an important process on the back walls of **cirques**, and also occurs on the downvalley side of a **roche moutonnée**.

plug SEE **volcanic plug**

plunge pool The pool at the base of a waterfall, which is formed by the erosive action of the rocks and boulders caught in the water. The force of the water and the erosion means that the pool is often deep and may undercut the rock wall behind the waterfall.

plutonic rock An **igneous** rock which has formed beneath the surface. Molten **magma** will have cooled slowly, and there will have been time for large crystals to form. All plutonic rocks are crystalline and coarse grained. The commonest plutonic igneous rock is **granite**, which is light in color because its mineral content is acidic. The commonest of the basic plutonic rocks is **gabbro**. Most plutonic rocks will have been formed in **batholiths** beneath the surface, but because of the erosion of overlying rocks, they are now commonly found on the surface.

podsol SEE **podzol**

podzol The commonest type of acidic soil, often occurring in areas of **coniferous forest**, where the needles fall to the ground and add to the acidity. The word "podzol" is Russian in origin and means ashy gray, which is a common color of podzols. They are leached soils, and the iron and lime which have been leached from the **A horizon** are often deposited as a **hardpan** layer in the **B horizon**. Large areas of podzol are found in the **taiga** regions of northern Canada, Finland, Russia and Siberia. They are normally quite poor agricultural soils, but can be improved by the addition of fertilizer. ALSO CALLED **podsol**

polar front The meeting point between cold polar air moving southward and warm tropical air heading northward. It is commonly located in the western part of the Atlantic near to the northern USA. As the two **air masses** pass each other, a whirling mass of air can develop, and in this way depressions are initiated. Once a **depression** has fully developed, the polar front will have been swamped by the whirling mass of air, which will then begin to travel eastward across the Atlantic Ocean.

polar winds Cold air which blows from the Arctic or Antarctic. The high-pressure region near the North Pole is a source of outgoing winds. They form large air masses, which travel southward. Polar **air masses** from Canada, called "Arctic highs," bring cold winters to the USA.

polder An area of flat low-lying land which has been reclaimed from the sea, especially in the Netherlands. The land is surrounded by embankments, called **dikes**, to keep the sea out, as many polders are below sea level. Polders need draining by pumping the water out into the ditches, which are numerous in the low-lying parts of the Netherlands. They are very fertile and can be used for growing **cereals** or bulbs, though much of the polderland is still left as rich pasture for feeding dairy cattle. There are small areas of polders in neighboring parts of Belgium and Germany.

pole l The North Pole and the South Pole are the extremities of the Earth's axis, around which the Earth rotates. **2** SEE **growth pole**

political geography One part of **human geography** which studies, among other things, the effect of geographical factors on political activities and problems, the effect of political actions on social and economic conditions, and the boundaries, extent and organization of political areas, especially countries.

pollution Contamination of the atmosphere or water, especially by chemical or industrial waste. Air pollution affects the **atmosphere**, and subsequently the land and oceans. Water pollution affects rivers, lakes and the sea. In the past it was quite common for industries and urban areas to use rivers and the sea as dumping grounds for their waste. There is now so much pollution that it is no longer possible to get rid of waste in this way, and it is known that some of the pollutant material is positively harmful. Much environmental damage has been done in the past through ignorance or carelessness, and there are now increasingly strict controls to try to prevent further damage. Clean Air Acts have reduced the amount of smoke, and in London, England, for example, **smogs** have virtually disappeared. They still occur in some locations, such as Los Angeles in the USA, where cars are the principal cause of the pollution. There is also a great deal of pollution in the atmosphere, and power stations seem to be a major cause. Many countries are now cleaning up their thermal power stations in order to reduce atmospheric pollution. It is believed that **acid rain** has been caused by industrial pollutants which have been emitted into the atmosphere and then brought down to Earth in rain. Many lakes have lost their fish; millions of coniferous trees have died, and it is thought that pollution has been the cause. Even stricter controls are necessary, and most developed nations are deeply concerned about the seriousness of the problem. **Third World** countries have less pollution at present because they have less industry, though in some localized areas they too have serious pollution problems. One of the most seriously contaminated and poisoned areas is at Cubatão, near São Paulo in Brazil. There is increasing anxiety about pollution of water supplies in North America and Western Europe because of agricultural fertilizers which have been washed into streams, where they can be harmful to animal life and possibly to humans. There is still much to be learned about the harmful effects of chemicals used by farmers, as well as by industrialists.

population l The total number of inhabitants of a city, country or region. In most countries, the population is surveyed every five or ten years. **2** (*in statistics*) The total number of anything, whether people, cities or trees in a park.

population change The increase or decrease of population in a country, region or city. The major reasons for change are variations in the **birth rate** and **death rate**, and in some areas **migration** will also be important.

population density The number of people per square mile or square kilometer. The figure can vary from 2 or less in parts of northern Canada, interior Australia or the Amazon basin in South America, to over 13,000 per sq mile [over 5,000 p er sq km] in parts of Singapore and Hong Kong. Large Western cities, such as New York, USA, or London, England, commonly have population densities of around 1,300 per square mile [500 per square kilometer].

population distribution The way in which a population is dispersed. It can be considered on a variety of scales. The world as a whole has many densely populated areas, notably parts of China and Japan, much of Java in Indonesia, and many parts of Western Europe and northeast USA. The world's sparsely populated areas include Australia, Canada, the Sahara Desert and Antarctica. On a smaller scale, in Europe there are many well populated areas, including southeast England, as well as sparsely populated areas, such as northern Norway and much of northern Scotland.

population explosion A large and rapid increase in the size of a population. With the help of medicines, chemical sprays and money from more advanced countries, many of the **Third World** countries experienced a dramatic reduction in their **death rates** during the 1950s, but there was no accompanying reduction in the **birth rate**. Previously the high birth rate had been matched by a high death rate, which ensured that population growth was quite slow. With this sudden change in circumstances, the population grew rapidly in what is now described as a population explosion. In the more advanced countries of the West, the birth and death rates had been changing gradually over the previous hundred years or so, but the Third World went through a drop in death rates from 40 per thousand to less than 20 per thousand in only a few years. The corresponding change in birth rate comes more slowly with education and improving economic standards, and some countries have not managed to achieve much progress. Birth rates in 1999 in Kenya, for instance, were still 30 per thousand, compared with a figure of 14 for the USA and 11 for Canada. The rapid increase of world population during the 1960s and 1970s was sometimes said to be exponential, and shows up very clearly on world population graphs, and in the **demographic transition**.

population growth An increase in the population of a city, county or country. The growth is the difference between births and deaths, with the added influence of **migration** in some cases. The world population has grown at a varying rate in the past, and the average at present is 1.3% per annum; in the 1970s a figure of 2% was normal. The world's population had grown to 1,000 million by 1820, and reached 2,000 million by 1930. A figure of 3,000 million was reached by 1960, and 4,000 million by 1975. The total population reached 6 billion in 1999, but the rate is slowing down. The figure is estimated to reach 8.9 billion by 2050 and is not expected to level out until 2200, when it is should peak at around 11 billion.

LARGEST CITIES (Estimated populations in 2015, in millions)		
Rank	City	No. of inhabitants
1	Tokyo, Japan	28.7
2	Bombay (Mumbai), India	27.4
3	Lagos, Africa	24.1
4	Shanghai, China	23.2
5	Jakarta, Indonesia	21.5
6	São Paulo, Brazil	21.0
7	Karachi, Pakistan	20.6
8	Beijing (Peking), China	19.6
9	Dhaka, Bangladesh	19.2
10	Mexico City, Mexico	19.1
11	Calcutta (Kolkata), India	17.6
12	Delhi, India	17.5
13	New York City, USA	17.4
14	Tianjin, China	17.1
15	Metro Manila, Philippines	14.9
16	Cairo, Egypt	14.7
17	Los Angeles, USA	14.5
18	Seoul, South Korea	13.1
19	Buenos Aires, Argentina	12.5
20	Istanbul, Turkey	12.1

population pyramid SEE **age-sex pyramid**

porous The term used to denote rocks containing pores (very small holes), which retain water. Sandstones contain from 5% to 15% pore space, and loose gravels have as much as 45% pore space. Most porous rocks are **permeable**, except for clay, which has such small pore spaces that they are blocked or sealed by water held in place by surface tension.

position The location of a settlement, in general terms, as opposed to specific detail. For example, the original site of London, England, was on an area of slightly higher and drier ground in a predominantly marshy

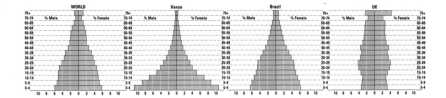

plain near the River Thames. Its position is in southeast England, at the lowest bridging point of the Thames. ALSO CALLED **situation**

postindustrial A term used to denote any of the more advanced nations which have seen a decline in the old heavy industries, notably steel, but also shipbuilding and often textiles. The older industrial nations have developed many new industries, mainly connected with electronics, computers or automation, and have a very large service sector of industry. The USA, UK and Germany are good examples.

pothole 1 A small, roughly circular hole on the rocky bed of a stream. Only a few inches in depth and diameter, it will be the result of **corrasion**, as small pebbles have been whirled around and scraped on the bed and sides of a small indentation. **2** A larger, vertical hole in an area of limestone, occurring where a **joint** has been enlarged by **solution**. If a pothole is used by a stream, it is called a **swallow hole**. Potholes generally extend from one **bedding plane** to the next, and along the bedding planes there may be caverns or channels eroded by flowing water. Potholers can progress up and down using potholes, and move horizontally by means of the river-formed channels. Many of the channels have become dry because the river has changed its route, probably to a lower level.

poverty cycle The vicious circle of poverty that is passed from one generation to the next, if it is not broken by outside factors (such as aid or training). In the cycle, poor parents cannot afford or do not want to send their children to school, leaving them with few or no qualifications,

▲ **population growth** The shape of an age-sex pyramid gives an indication of population growth and potential population growth. The relative numbers of young and old people reflect relative birth and death rates. More importantly, however, a predominantly young population will tend to produce more children in the future. World population growth is rapid. From the example pyramids it can be seen that population growth in Brazil will be about average for the world, in Britain population will grow very little, and in Kenya it is likely to grow very rapidly indeed.

meaning they may only find poorly paid work, which leaves them unable to afford to bring up their own children in better conditions.

prairie A flat grassland area. The prairies are the midlatitude **temperate** grasslands of central Canada and the USA, and extend from Alberta, Saskatchewan and Manitoba in Canada, southward through North Dakota and South Dakota, to Kansas, Oklahoma and Texas. They are flat and treeless areas, and the climate is ideal for grass to grow. There is summer rainfall, with about 30 in [700 mm] per annum in the east but decreasing westward to only 12 in [300 mm], and the winters are fairly dry, with just a little snow in the north. Summer temperatures are 70°F [20°C] or more, but winters can be very cold in the northern parts; temperatures of 15°F to −5°F [−10° to −20°C] are often experienced. Large areas of the prairies have been plowed for **cereal** production, and the wetter eastern parts are very productive. The drier western prairies, where the grass is shorter, are less productive, and repeated plowing in several places has given rise to **dust bowl** conditions. Before the plowing took place, the grasslands were the home of large numbers of cattle and sheep, which were looked after by cowboys. Prior to the European settlement, the prairie landscape had been the home of millions of buffalo and numerous groups of Native Americans. The **steppes** of Russia are very similar to the North American prairies, and in the southern hemisphere there are also temperate grasslands: the **pampas** in South America, the **veld** in South Africa, the Canterbury Plains in New Zealand, and the Murray-Darling basin in Australia, but they are milder and wetter than those of the northern hemisphere.

Pre-Cambrian The earliest geological era, which finished about 570 million years ago. Nearly one quarter of the world's land surface is covered by Pre-Cambrian rocks, but very little is known about them because they have undergone so much erosion, folding and faulting, and most have been metamorphosed. Also, there is very little fossil evidence. Because only very primitive forms of life existed during that period, it is more difficult to work out the paleogeographic conditions of the time than in the case of more recent rocks. Pre-Cambrian rocks occur in the Grand Canyon in the USA and over much of Canada, large parts of southern Africa, Western Australia and large areas of Siberia. The Baltic Shield in Finland and neighboring countries is the largest European area of Pre-Cambrian rocks.

precipitation The products of condensation in the **atmosphere** which fall on the surface of the Earth, including rain, snow, hail, sleet, dew and hoar frost. They can be measured in a rain gauge, but the snow and frost has to be melted first. Total precipitation is recorded in inches or millimeters.

preindustrial A term used to denote an early, mainly rural society in which agriculture was the predominant activity. The preindustrial societies were originally subsistent, but gradually trade developed as communications improved. The use of metals and the development of craft industries contributed to the growth of towns and the development of nonagricultural settlements. Industrialization began with the use of steam power at the end of the 18th century, after which many towns grew rapidly, and an urban society evolved. Industrialization took place at different times in different countries, as outlined in **Rostow's model**.

pressure gradient The change of pressure as shown by **isobars** on a weather map. The isobars are similar to contours, and the pressure gradient can be compared to a hill. The closer the isobars are, the steeper is the hill and the stronger the wind. The presence of few isobars will indicate weak winds.

prevailing wind The most frequent wind direction. Some parts of the world have very persistent winds; for example, the **trade winds**. In temperate latitudes the winds are more variable, especially where **depressions** occur. In depressions the winds blow from all directions, and so they do not bring dominant winds.

primary era SEE **Paleozoic**

primary industry Any of the industries which provide primary products, that is, goods which have not been changed or processed. The primary industries include mining and quarrying, forestry, agriculture and fishing, and together they constitute the **primary sector** of the economy.

primary sector The part of the economy which is concerned with the **primary industries**. Processing and manufacturing is regarded as the secondary sector; the tertiary includes all the services, while quaternary industries form the information and expertise sector. The percentage of working people employed in each of these sectors is a guide to the stage of economic development reached by a country. **Third World** countries are still very much involved in agriculture, in the same way that, for example, the USA, Canada and European countries were 300 years ago. Over 90% of the working population in some countries, such as Mali or Nepal are involved in

▶ **precipitation**
Shallow clouds and those in the tropics do not reach freezing level, so ice crystals do not form (A). Instead, a larger-than-average cloud droplet may coalesce with several million other cloud droplets to reach raindrop size. Electrical charges may encourage coalescence if droplets have opposite charges. Some raindrops then break apart to produce other droplets in a chain reaction that produces an avalanche of raindrops. Most rainfall in the mid-latitudes is the result of snowflakes melting as they fall (B). It takes many millions of moisture droplets and ice crystals to make a single raindrop or snowflake heavy enough to fall from a cloud. Yet a snowflake can be grown from ice crystals in only 20 minutes. Large

hailstones need strong upcurrents of air in order to form (C). A 30-mm (1.2-in) diameter hailstone probably needs an updraught of 100km/h (60mph). The turbulent air currents in a thunderstorm turn a frozen water droplet into an embryonic hailstone. The abundant supercooled moisture

droplets in a storm will readily freeze on to its surface. It is swept up and down by the currents and accumulates numerous layers of thick ice, which are alternately

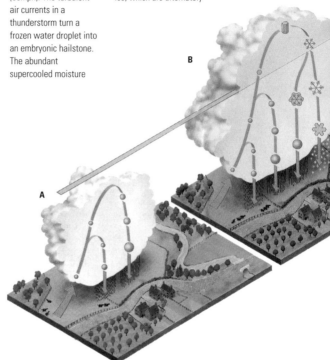

primary activities. In addition to national differences, there are also regional differences in employment. SEE ALSO **secondary sector**, **tertiary sector**, **quaternary sector**

primate city The largest city in a region or country, often the capital city, such as London or Paris. In some countries, such as the USA or Brazil, the capital is not the largest city. London and Paris are much larger than the second cities in size in England and France, and they are said to have a high degree of primacy.

primeur *(French)* A term denoting early fruit and vegetables. If early produce can

be taken to customers before rival producers have grown their fruit and vegetables, a higher price will be obtained. In the USA, it is Florida and California that control the early season crops. In France, the southern Mediterranean areas are able to grow tomatoes, strawberries and flowers for Paris and other markets much earlier than they can be produced in the north of France. In the UK, the Isles of Scilly are able to grow daffodils earlier than anywhere else. Nowadays early produce can be grown in **greenhouses**, using artificial heating, but this makes it more expensive. In Iceland, greenhouses grow bananas, peppers and other products using **geo-**

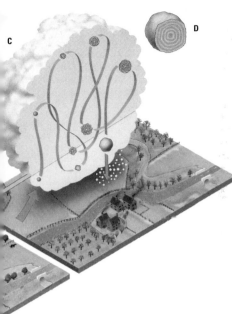

obtaining oil from olives, juice from oranges, cheese from milk and wine from grapes. Processing may involve purification or refining.

profile I A vertical section of soil, which shows the changes in the soil, including the different horizons from the surface down. SEE ALSO **A horizon, B horizon, C horizon. 2** A drawing of a section of a river valley. SEE ALSO **river profile**

promontory A headland or small peninsula which protrudes into the sea. It will suffer from **erosion** along both sides, and is likely to have cliffs and possibly caves as a result of **wave** action.

public sector That part of a national economy which is owned and controlled by the state. For example, central and local government and nationalized industries.

pull factor SEE **migration**

pumice A type of volcanic rock which is very light and porous. It contains a large number of holes, which are the result of escaping steam and gases at the time of its formation. It is so light it will float, and it tends to be very acidic in chemical content. Much of it originates as a kind of froth or scum floating around on top of the lava.

pumped storage A system employed at some hydroelectricity stations, whereby water is pumped back uphill to a storage reservoir during the night, when surplus electricity is available. The water can be reused the next day to generate electricity, and it helps power stations to operate with quite small storage reservoirs.

puna An area of high, level and bleak land in the Andes, between 10,000 ft to 13,000 ft [3,000 m to 4,000 m] above sea level. Temperatures can be low, even in mid-summer, as the atmosphere is rarified because of the altitude. Summer days may reach 50°F [10°C], but the night temperatures fall well below freezing point. Throughout the year, bitterly cold winds

clear and milky. The opaque layer is made when air bubbles and sometimes ice crystals are trapped during rapid freezing in the cloud's cold upper levels. The clear layers form in the cloud's warmer, lower levels, where water freezes slowly. There can be as many as 25 layers in a hailstone (D) and the last – a clear layer of ice, which is often the thickest develops as the hailstone falls through the wet, warm cloud base.

thermal heating, provided by hot water from underground.

private sector That part of a national economy which is not owned or controlled by the state. Examples include labor unions and commercial companies.

privatization The sale of a formerly nationalized, or state-owned, industry or property to the public. It is also known as "denationalization."

processing industry An industry which converts an agricultural commodity into a saleable product. Examples include are common. The local Indians sometimes grow hardy cereals and potatoes, and rear llamas and alpacas. Life is very hard and the people are poor. They keep warm by staying in the sun for as long as possible, but they wear woolen ponchos made with wool from their own animals. There are a few rich mineral deposits on the puna of both Peru and Bolivia, and the local people provide a source of low-paid labor, the profits going mostly to a few rich landowners.

push factor SEE **migration**

puy A **volcanic plug**, taking its name from Le Puy, a town in the Massif Central in France, where there are several examples, including one which has a monastery built on its narrow summit.

pygmy A race of small people, who grow to about 4 ft 5 in to 4 ft 7 in [135 cm to 140 cm]. The most famous groups are the Negrillos of the central African forests, but there are others in Southeast Asia, who are known as Negritos. They are mostly semi-nomadic hunters and gatherers who live in tropical forest areas. They are decreasing in numbers because of outside influences and many of them have become settled farmers or shifting cultivators.

pyramidal peak SEE **horn**

pyrites Any of various yellow minerals that have a metallic luster, especially iron pyrites (FeS_2), which is a source of iron and sulfur. Sometimes the color of pyrites has led to its wrongful identification as gold, and it is often referred to as "fool's gold." Copper pyrites is similar to iron pyrites, but has a different mineral content. It is found in some **igneous** rocks, and often occurs in veins or seams of igneous materials which have forced their way through existing rocks in liquid or gaseous form. When they solidify, they may be mined or quarried for a variety of minerals, not only pyrites.

PROFILE – SOIL

Shown here is one example of a soil profile, but there are many others. Soil profiles are used in classifying soils. The composition, color, and order of layers identified a soil type to a pedologist. For example, a tundra soil will have a dark, peaty surface, whereas the top layer in a desert soil will be light-colored, coarse and poor in organic matter. One layer of soil in a soil profile is known as a "soil horizon."

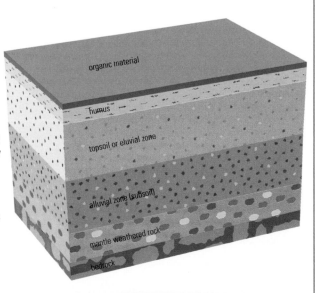

organic material

humus

topsoil or eluvial zone

alluvial zone (subsoil)

mantle weathered rock

bedrock

pyroclastic rock A fragment of rock thrown out by volcanic activity, generally in association with some violent explosive action. Pyroclasts normally include solidified lava left behind by a previous eruption of the volcano, as well as rocks from the crust, smaller pieces of cinders, ash and dust. The largest pyroclastic fragments may weigh several tons, and they are called "volcanic bombs." Smaller pieces are called "lapilli." When pyroclastic activity ceases, outpourings of lava often follow.

Q

qanat An underground irrigation channel which brings water from the foothills to the neighboring plains. There are examples in Pakistan and Iran. ALSO CALLED **kanat, karez**

quadrat An area of land, generally square in shape, in which a field study is taking place. Quadrats are used most frequently for vegetation surveys in ecological studies; the most popular size is 10 sq ft [or 1 sq m]. If trees were the major subject of the study, a quadrat of 100 sq ft [or 10 sq m] would be more suitable. Within the area of the quadrat, a detailed study of the plants can be made; it will be regarded as a sample study, which can be compared with other quadrats studied elsewhere. On a **transect** line across a valley, for example, it would be possible to take a quadrat sample of vegetation every 10 ft or 100 ft [or 10 m or 100 m], depending on the scale of study and the size of the quadrats.

quality of life The standard of living together with other, less definable considerations, such as the amount of noise and pollution, the availability of countryside, flowers and birdsong, etc. Conservationists in particular are very concerned about such considerations, which are not always available to people living in large cities, although they will have museums, theaters, movie theatres, etc., which are also factors to be considered. Geographers have to try to consider all aspects when assessing problems such as conflict over land use.

quantitative Considering quantities. Geographical studies became increasingly quantitative during the 1960s and 1970s, and the purely descriptive qualitative work became less important, although still an essential part of geographical studies. Instead of describing slopes, for example, geographers went out and measured them and then tried to explain how they had formed and why they were different from other slopes. Quantification meant that problems were solved by examining data and trying to be objective rather than subjective.

quarrying The excavation of rock from an open working on the surface. Many building materials, such as limestone, sandstones and granite, are quarried. Chalk is also quarried for use in cement and lime. Occasionally coal and iron ore are extracted in this way from open-pit workings. Quarries sometimes create conservation problems, as they can become eyesores. Many old quarries cannot be changed now, but in modern workings great care has to be taken of the landscape, and whenever possible trees are planted to hide the workings.

quartile One of four equal parts in a data distribution set.

quartz A very common mineral, found in many of the Earth's rocks. It consists of silica; that is, silicon and oxygen. Most grains of sand consist largely of quartz, and it is very common in **granites** and other acidic igneous rocks. Pure quartz is colorless and glassy, and it is the major source of glass. It is a hard mineral, the hardest of all the common substances.

quartzite A metamorphosed sandstone which is very hard and resistant. It is found, for example, thoughout Nevada, USA, and in the La Cloche hills in Ontario, Canada. It consists mainly of quartz and is cemented by silica. There are also some softer quartzites, which are sedimentary in origin.

Quaternary A geological period representing the last 2 million years. It consists of the **Pleistocene**, which was the **Ice Age**, and the **Holocene,** which is the time after the Ice Age. During the Quaternary period, there were several ice advances, followed by retreats during which interglacial conditions prevailed. These were sometimes warmer, but sometimes cooler than the present climatic period. The present is thought likely to be an interglacial period, rather than definitely postglacial. The glacial periods were similar to the conditions found in Greenland and Antarctica at the present time. The interglacials included some **tundra** conditions with **periglacial** processes being active, as well as some warmer periods when deciduous forests were able to grow.

quaternary sector The part of the economy which deals with providing information and expertise. The quaternary sector has grown rapidly in recent years. Quaternary sector industries are found mostly in advanced countries, especially near main centers of modern industrial development. Such activities might include law, research or finance. SEE ALSO **primary sector, secondary sector, tertiary sector**

quicksand An area of sand and mud which contains a high proportion of water. It generally occurs near a river or the sea, where the **water table** is high, but **drainage** is poor. It is because quicksand is almost liquid that animals may sink into it.

quota A fixed maximum amount of a good that may be imported, exported or produced. Such levels are usually set by governments or international organizations to limit quantities of particular goods, especially if there is overproduction.

R

race A classification, usually used broadly to describe a group of distinct humans. The people in each group have identifiable characteristics marking them out; for example, facial features or skin color. Since the world population today is very mixed through migration and interbreeding, it is not common to find all members of one race living only in one place.

radial drainage A pattern of rivers in which the flow of water is away from a central high point. The rivers all radiate outward, like the spokes of a wheel. The central high point is likely to have been a dome originally, although erosion over millions of years will have changed its shape. SEE ALSO **drainage**

radiation The emission of energy in the form of heat. Radiant energy is given off by the sun, and some reaches the Earth, where it provides heat. The Earth gains heat from radiation from the sun, but also loses heat by radiation from the ground. During the day, more heat is received from the sun, by **insolation**, than is lost by the ground. At night, the loss is much greater. There are large differences between the seasons, because the long days in summer enable the ground to warm up, whereas in winter it will become progressively colder. Land loses and gains heat more rapidly than water, which contributes to the climatic differences between the continents and the oceans.

radiation fog Fog which occurs as a result of heat loss from the ground by radiation during the night. Clear skies and calm weather conditions are ideal for the formation of radiation fog. SEE ALSO **fog**

radioactive decay The breakdown of certain unstable particles in rocks which enables the age of the rocks to be calculated. The rate of change is based on the half-life method of dating. This means that half the radioactivity is lost after a set time – say, 20,000 years. After a further 20,000 years there will be another loss of 50%, and so on, since there is a gradual decrease in radioactive content. Each mineral has its own half-life period. The radioactive isotopes change as they decay, from "parent" to "daughter" isotope: uranium-238 (U^{238}) changes to lead-206 (Pb^{206}); uranium-235 (U^{235}) to lead-207 (Pb^{207}); potassium-40 (K^{40}) to argon-40 (Ar^{40}).

radiocarbon dating A method of dating using the radioactive isotope of carbon called C^{14}. Carbon-14 oxidizes to carbon dioxide. It enters the Earth's carbon cycle and is absorbed by plants. Once the plants have died, the carbon-14 diminishes at a known rate. The rate of decay is known, because carbon-14 has a half-life of 5,570 years. The age of wood or peat, as well as bones or shells, can be dated by analyzing the proportion of radiocarbon in the total amount of carbon that is contained in the material. The radiocarbon method can be used for measuring ages of up to about 70,000 years, although above 25,000 to 30,000 years it becomes less accurate.

rain Drops of water falling from the sky. They are formed by condensation of water vapor in the **atmosphere**, caused by cooling of the air. Small droplets form cloud, but as the drops grow they become too heavy to float, and so they descend to the ground. They generally have to exceed 0.02 in [0.5 mm] in diameter in order to overcome the effects of slowly rising movements of air.

rainbow A bright, multicolored band, usually seen as a circular or partial arc formed opposite to the Sun or other light source. The primary bow is the one usually seen; in it the colors are arranged from red at the top to violet beneath. A secondary bow, in which the order of the colors is

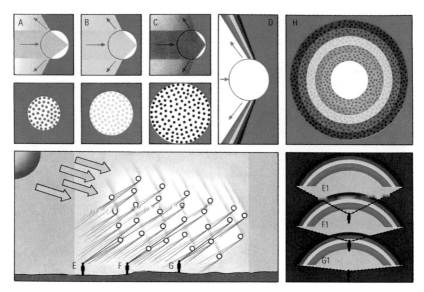

▲ **rainbow** A rainbow is a natural demonstration of the mixture of wavelengths that makes up sunlight. Drops of moisture in the atmosphere act as prisms, dispersing the light into its component colors. Violet (A) is always refracted, diffracted and finally reflected through raindrops at an angle of about 40° parallel to the Sun's rays, yellow (B) at about 41° and red at about 42°, with other colors at angles between these. In this way the complete spectrum is formed (D). As each color is formed by rays that reach the observer at a certain angle, no matter what the position of the observer (E, F and G) in relation to the Sun and the raindrops, the same spectrum is seen as E1, F1 and G1. A rainbow is essentially a circle (H), but part of the circle of light in a rainbow usually reaches the ground before it reaches the observer so the observer sees only a section of an arc. From an aircraft, however, an observer may see the complete circle.

reversed, is sometimes seen beyond the primary bow. The colours are caused by reflection of light within spherical drops of falling rain, which disperse white light into its constituent wavelengths. The colors usually seen are those of the visible spectrum: red, orange, yellow, green, blue, indigo and violet.

rain day A period of 24 hours in which a measurable amount of rain has fallen. The minimum measurable amount is generally taken to be 0.01 in [0.25 mm].

rainfall 1 A shower or fall of rain. All rain is formed by rising air, which is then cooled to enable condensation to take place. The air rises for a variety of reasons and so there are different types of rainfall, including convectional, cyclonic, frontal and orographic rainfall. **2** The amount of precipitation received by a given area over a given period.

rain forest A dense forest growing in regions with heavy rainfall throughout the year. Such forests are most numerous in tropical latitudes, such as the Amazon basin in South America, Zaïre in south-central Africa, and parts of Southeast Asia. There are examples in warm temperate regions such as eastern Brazil, Kwa Zulu-Natal in South Africa, and parts of China and southern Japan. There are also examples in wet maritime regions, such as Washington state in the USA, Tasmania, southern New Zealand and southern Chile.

rain gauge An instrument used for measuring rainfall and other forms of pre-

RAIN GAUGE

The rain gauge illustrated is designed for daily measurement of total rainfall. The outer case (1) is shaped to provide stability when the gauge is embedded in the soil. The gauge is essentially a funnel (2) of standard diameter to direct the rain to the inner can (3). The rainwater is poured into a special measuring cylinder (4) graduated to read inches or millimeters of rainfall.

cipitation. There are many different types, including some self-recording varieties. They consist of a cylinder which stands up above the ground surface to prevent splashing water bouncing in. Inside the cylinder there is a funnel to lead the water into a collecting jar. The water collected is poured into a measuring cylinder, and the total rainfall can be measured off, generally in inches or millimeters. The commonest rain gauges have a diameter of about 5 in [130 mm], and will be read every 24 hours.

rain shadow SEE **orographic rainfall**

rainwash A process in which drops of rainwater hit rock and soil, washing away tiny particles. It is especially effective in areas with no vegetation cover, and can be a cause of soil erosion. Rainwash is a major contributor to the erosion of valley sides, and helps to create the V-shaped cross-profile of valleys in humid regions. The lack of rainwash in deserts is a reason for the steep sides of **wadis** and **canyons**.

raised beach An elevated beach, which is now above sea level because of an uplift of the land or a retreat of the sea. It is often bounded on its inland edge by old cliffs, sometimes with caves. The uplift of land may have been the result of tectonic movements, or may be the result of isostatic re-adjustment of the land after ice has melted. When the weight of ice has been removed by melting, the land gradually moves up, relative to the sea. Raised beaches of up to 200 ft [60 m] have been formed in Scotland because of isostatic movement, and some of over 650 ft [200 m] are found in Norway. The same effect can be produced by the retreat of the sea, and this occurred when snow and ice increased during the glacial phases. More snow and ice meant less water in the sea, and so the sea level went down. Some locations have a series of two or three raised beaches because of the several changes of sea level during the **Ice Age**.

rakes SEE **lode**

ranching A type of pastoral farming, generally found in grassland areas. Large-scale and extensive methods are used to rear animals, usually cattle, but also sheep. Formerly, the animals could roam around over large areas and were looked after by cowboys, but nowadays there are more fences to restrict the movements of the animals. Cowboys still exist in some places such as the estern USA and the Northern Territory in Australia, but vehicles are often used instead of horses. Argentina, Uruguay and parts of Australia are major areas for ranching. Many of the animals are fed on alfalfa, corn and other feedstuffs, as well as being left to fend for themselves on the natural pasture.

random Without a regular pattern. A haphazard arrangement with an equal chance of any event or location occurring. In statistics, it is applied to numbers in a set that are just as likely to come up as any others.

Randstad An urban area in the Netherlands which includes Amsterdam, Rotterdam, The Hague, Dordecht and Utrecht. It is a ring of almost continuous urban development, with a **green belt** in the middle of the ring. It is the original and best example of a **ring city**.

range 1 A line of mountains which makes a more or less continuous barrier, even though there will be passes or gaps through it. The Rockies or Appalachians in the USA are good examples. **2** An open area of grazing, as found in the high plains of the western USA. **3** The difference between the highest and lowest temperatures experienced in any given period, such as a day or year. **4** The difference between the high and low points of a tide. The range will be greater at the time of spring tides than at **neap tides**. **5** The distance over which a commodity will be distributed from a central place. This depends partly on the maximum distance a consumer is willing to travel in order to purchase the commodity, which in turn depends on how frequently it is required. For instance, low-order goods with a daily requirement, such as bread or newspapers, will be obtained from somewhere close by. High-order and more specialized goods which are only required every few weeks or months may be obtained from many miles away.

ranking The placing of a set of data in numerical order so that the highest figure is ranked one, the next highest is ranked two, and so on. Ranked information is used in many statistical techniques, especially for correlation.

rank-size rule The theory that the population of the second largest city in a region or country will be half the population of the largest city of that region or country. The population of the third largest city will be one-third the size of the largest city, and

▼ **raised beach** As the level of water changes in relation to the level of the land, the original beach or wave-cut platform is left above the highest water level as a raised beach. The original cliff-face usually degrades because the action of the sea no longer cuts into the base of the cliff.

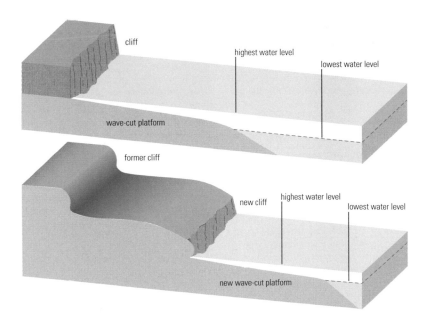

the fourth city will have a population one-quarter the size of the largest city. The formula employed in the rank size rule is:

$$Pn = P1\ (n)^{-1}$$

where Pn is the population size of a city of rank "n," and P1 is the population of the largest cith in the area. The rank-size rule fits many countries, especially if industrialization has taken place. In some countries, such as the UK or France, the excessive size of the primate city makes all other cities look incorrect according to the rank-size rule. However, it is really the large size of London and Paris which upsets the theory in these two countries.

rapids Small waterfalls and turbulent water along a stretch of river. They are generally the result of a different and harder type of rock outcropping on the bed of a river. Rapids are gradually worn down by erosion, and eventually disappear. Steep rapids, such as those on the upper reaches of the River Nile in Africa, are called "cataracts."

ravine A small, narrow, steep-sided valley.

raw materials A primary or partly processed product which is the basis for manufacturing into a finished product. Raw materials are the necessary requirements of industries. Examples include agricultural products, minerals and timber, as well as many half-made goods such as cotton which has been spun but not woven, wood pulp, or flour.

recessional moraine SEE **moraine**

reclaimed land Land which is recovered for productive use. It may be reclaimed from the sea, from marshes, or from dereliction after spoiling by mining, quarrying or industry. "**Polder**" is the name given to land reclaimed from the sea in the Netherlands, and the Fens are the best example in Britain. Barriers called embankments or **dikes** are built to keep out the sea water, and pumping is necessary to keep the land dry enough for crops to be grown. Often grassland is grown on the reclaimed land and cattle can be reared. In Languedoc in southern France, the land was marshy and malarial until the 1950s, but has now been successfully reclaimed for tourism developments. Marshes near Naples in southern Italy and Catania in Sicily have been reclaimed, and are used for growing fruit and vegetables.

recreation A pastime, diversion or other leisure activity. Recreation is a growth area, both for vactioners who are away from home, and for local people who may be looking for recreation in an amusement park, theater, sports center or other recreational facility. Resorts which cater for sun-seekers or skiers are a major branch of the recreation industry which has grown rapidly in the last 30 years, with the development of air travel, package tours, longer vacations and more money.

recycling The process of collecting and recovering waste for reuse. Many minerals can be collected as scrap, melted down and reused. Recycling is already commonly practiced in steel manufacture, but can also be applied to lead, copper and tin. There are aluminum collecting centers in some places, but most forms of reusing metals are uneconomic at present. However, in order to conserve the diminishing supplies of metals, it will be vital for recycling to become more common. Recycling of bottles is also growing in importance, and there are several places where paper is recycled. Technology has been slow to develop equipment for re-using many of the raw materials which are exhaustible, because it has always been expensive and there was no great need in the past. Now that it is realized that many of the Earth's resources will be exhausted during the 21st century, there is increasing pressure on industry to reuse materials and be less extravagant than in the past. This will inevitably add to the cost of many items.

red clay SEE **pelagic**

red rain Rain coloured by fine dust particles that falls in midlatitudes. The dust is picked up in dry areas and carried by wind to more humid areas, where it falls in rain. For example, dust from the Sahara Desert falls in rain over southern Europe.

reentrant A small valley or lowland area extending into a region of higher ground. It will have been formed by water erosion or by **freeze-thaw** and **solifluxion**.

refining The process of removing impurities from metals, sugar, petroleum, etc. For metal ores, especially, it is best to carry out the refining process near the mines in order to avoid unnecessary transportation costs. Iron ore is from 25% to 70% pure, and a low-quality ore of only 25% contains 75 tons of unwanted rock in every 100 tons extracted. Whenever possible, unwanted minerals are separated out before transportation.

reforestation The planting of trees in an area which has been deforested. In many parts of the world, trees have been cut down, and the land has been used for agriculture, as is happening in Brazil today. In other places, such as Nepal, trees are cut down on the slopes of the Himalayas, and the exposed soil is immediately washed away. It is very important that reforestation should take place in such areas in order to prevent **soil erosion**. In other forested areas, such as Sweden and Canada, large numbers of trees are cut down for timber and pulp, but they now plant as many trees as they cut down. Reforestation is important in these areas, too. SEE ALSO **afforestation**

refugee A person seeking shelter and protection in a foreign country due to persecution or troubles (such as war) in their own country. In 1995, there were estimated (by the United Nations) to be some 14.5 million refugees in the world, a number which is still rising rapidly. Refugees will often receive only temporary accommodation in the countries to which they go, and are usually not accepted fully into the society, lacking any rights or power.

regelation The process of refreezing. In a glacier, a considerable amount of melting occurs because of the effects of pressure. The melting point of ice is lowered by the pressure caused by the weight of the ice. If the pressure is released, then the melting point will rise and the water will freeze again. Regelation is probably very important in explaining the movement of a glacier.

regime The seasonal changes in climatic conditions. The way in which temperatures and rainfall change from month to month. SEE ALSO **river regime**. **2** A political system.

region An area which has a number of distinctive characteristics. It may be very small, such as a tiny valley, or very large, such as the Amazon basin in South America. A region may be defined by one or several of its characteristics. Climate, relief, rock type, agricultural pattern, natural vegetation and many other features may be the reason for the delimitation of a region. The Sahara is a region because of its climate, and London is a region because of its urban development and buildings.

regional geography That part of geography studying regions and their particular qualities. In this study, it is especially important to define the regions and their boundaries very clearly, and it will include both physical and human features and the link between them. Regional geography was most popular in the 1950s but has declined in importance since then.

regional inequality The differences in standards of living existing within a country. Factors which show regional inequalities include unemployment rates, per capita income, numbers of people in higher education, housing standards, or types of industry in different areas. Governments then try to adapt regional policies to alleviate particular factors.

research and development (R & D)

Facilities found within most large industries devoted to researching new products or techniques and developing them to help the particular industry. Research and development facilities can also be found independently and through the government.

regolith   Loose rock fragments which have broken away from the solid **bedrock**. The fragments may be very thin and are sometimes patchy. Eventually, they will be broken down into soil particles, and sometimes the word "regolith" is considered to include soil and superficial deposits such as **alluvium**, glacial **drift** and **loess**.

regression line   A statistical technique, that shows a "best fit" line through a series of points on a graph and attempts to explain the relationship between two or more sets of variables, that may be dependent on, or independent of, each other.

Reilly's Law of Retail Gravitation

A model devised by the American W.J. Reilly to enable predictions to be made about shopping habits and the likely limits of **catchment areas** for malls and shopping centers. It uses the relationships between distance and population to provide the answers, and assumes that all people are logical in making their decisions about where to shop.

rejuvenation   A renewal of the erosive power of a river because of a change of **gradient** in the river's **long profile** or the arrival of extra water. Extra water may be the result of **river capture** or because of **meltwater** from a glacial area, as would have happened at the end of each glacial phase in the **Ice Age**. A change of gradient would be the result of a fall in sea level, or of a rise in land level near the source of the river. In any of these situations, there would be renewed vertical **corrasion** along the river bed. Several distinctive features are formed by rejuvenation. The **knick point** is the point at which increased erosion begins, and is simply a break in the long profile of the river, marked by turbulent water or a small waterfall. The knick point is gradually eroded and the break on the long profile will work its way upstream. Rejuvenation causes increased erosion, and some **meanders** may become incised. The river may cut down to a new and lower level, leaving the old flood plain abandoned at a higher level as a **river terrace**.

relative humidity   The amount of water vapor actually present in the air, expressed as a percentage of the maximum that could be held by air at a particular temperature. It is a measure of dampness, and is normally measured by a **hygrometer** using wet and dry bulb thermometers. The warmer the air, the more water vapor it can hold. It is possible, for example, for air on a dry day in the Sahara to be holding more water vapor than the air in New York, USA, when it is actually raining. If air is cooled, it can hold less water vapor; for this reason, dew is common during the night in the Sahara. It is also the reason for the formation of **cloud** and rain when air rises. Air at a temperature of 50°F can hold 0.009 oz per cu ft (at 10°C, 9.4 g per cu m); air at a temperature of 70°F can hold 0.017 oz per cu ft (at 20°C, 17.3 gm per cu m); air at a temperature of 90°F can hold 0.032 oz per cu ft (at 30°C, 30.3 gm per cu m).

relief   The features of the landscape. Differences in altitude, steep slopes and gentle slopes, flat plains and cliffs are all part of the relief of a landscape. The term includes only the physical features, unlike **topography** which also includes the man-made features of a landscape. The study of relief features is called **geomorphology**.

relief map   A map which shows relief features, by contours, or shading and colouring. Many Swiss maps use "hachuring," which is a system of lines drawn down the slopes. The steeper the slope, the more numerous the lines. Some maps use hill shading, in which hills are shaded on one side to give a pictorial impression of relief, as though a light is shining from one edge of the map and casting a shadow on the opposite sides of hills.

relief rainfall SEE **orographic rainfall**

remote sensing The use of spacecraft and satellites to take photographs from great heights. As well as conventional photographic equipment, various infrared and electronic devices are also used, which produce images that may need specialist interpretation. A common use of satellite photographs is in weather forecasting, because cloud height and thicknesses can be determined, as well as temperatures in the oceans. Remote sensing enables information to be gathered from inaccessible areas and the photographs can be used to show the changing vegetation patterns from season to season, or from year to year.

rendzina A soil type which occurs on chalk and limestones. The **A horizon** is often loamy though quite thin, and the **B**

horizon contains fragments of chalk or limestone rock. Many rendzinas are changing because they are being plowed and used for **cereal** production. Rendzinas are found in many regions, including parts of eastern Texas and Alabama in the USA.

renewable fuel Any form of energy of which there is an infinite supply. Hydroelectricity, wave power, wind power and tidal power are all renewable forms of energy, supplies of which will be available to future generations. At present, nuclear power is nonrenewable, though it is possible that scientists will develop a renewable method of creating energy from nuclear fuels.

renewable resource A resource which can be replenished, such as wind and water. Soil is a renewable resource if it is carefully

REMOTE SENSING

Remote sensing satellites (1) view the Earth from space using various sensors and cameras. These instruments are classed as either active or passive. Passive devices, such as optical cameras and infrared scanners, pick up reflected radiation (either light or heat). Active instruments on the other hand send out radio pulses and record the returning signal. One of the strengths of active scanning is the ability to see through cloud. Powered by a solar sail (2), the satellites use orbits that take them above the whole of the Earth during a series of days. Images of the Earth's surface are beamed down to ground stations (3) in digital form (4), and are converted into pictures by computers (5) so that they can be interpreted.

managed. Careless or thoughtless farming may destroy the soil and cause **soil erosion**, and in this case the soil cannot be regarded as a renewable resource. By replanting as many trees as are cut down, timber can be a renewable resource. Selective cutting enables forests to survive permanently, and in some places wholesale clearance is followed by immediate replanting.

representative fraction The scale of a map expressed as a fraction of the actual size of the area represented. For example, if a map has a scale of 1:100,000, this means that one unit on the map (such as an inch or a centimeter) represents 100,000 of the same units on the surface of the Earth. The smaller the second number, the larger the scale, so a 1:50,000 map can have more detail than a 1:100,000 map.

reservoir A lake used for storing water. Sometimes it may be a natural lake, but generally reservoirs are man-made lakes which have formed behind dams. They often provide hydroelectric power, as well as water for human or industrial consumption, and possibly for irrigation, too. Large reservoirs are found in many countries, often where there is a marked seasonal distribution of rainfall. The heavy rains of the wet season are stored for use in the dry season; for example, in Lake Nasser on the Nile, the Damodar valley reservoirs in India, or the reservoirs in the Snowy Mountains in Australia.

reservoir rock A porous rock in which the pores have been filled by water, natural gas or oil. If the rock contains water, it will be described as an **aquifer**. Reservoir rocks containing oil or gas need a trap to prevent the substance escaping; this often occurs in anticlinal formations. The oil or gas moves to the top of the **anticline**, but cannot escape because there are **impervious** layers above and below. Traps are also found near salt domes and fault lines. Reservoir rocks are sought by geologists when they are looking for oil or gas deposits.

residual deposit Weathered and eroded fragments that are left behind. **Weathering** occurs *in situ* and leaves fragments of rock on hillsides as **scree**, or on top of limestone or chalk as **terra rossa** or clay with flints. Erosion may involve some movement, and residual deposits may be dumped by ice or rivers.

residual hill A small but steep, isolated hill; the last remnant of a larger mass of high ground. It may be the middle of a compressed anticline, a metamorphosed area, or the **puy** in a volcano. SEE ALSO **monadnock**

resistant rock A relatively hard rock which can withstand **weathering** and **erosion** more successfully than neighboring rocks. Resistant rocks often form higher ground, as the softer neighboring rocks are worn away. They may be hard because of their mineral content, or possibly because the particles are well cemented, or because of the absence of **joints** and lines of weakness. Whether or not a rock is hard and resistant will depend on the adjacent rocks; resistance is relative.

resource management The careful and considered control of the exploitation of a resource, based on factors such as economic and environmental cost, future needs and national policy. It may be that it is decided not to use a resource at a particular time. Resource management therefore involves resource development and resource conservation.

resurgence The emergence of a stream from underground, usually after flowing through **permeable** rock strata and having come to an **impermeable** rock layer.

retailing The selling of goods. Retailing is a small-scale activity, in comparison with wholesaling, and is generally carried out from a store. The term "retailing" is sometimes extended to mean more than merely sales from stores, and can include such activities as hairdressing, banking, catering and the world of entertainment. Most retailing is found in downtown areas (the **central business district**). Larger settlements have a greater number of retail functions than can be found in a small settlement. Small towns may have only one or two retail outlets, a small city may have 50 to 100, while a large city may have hundreds or thousands. The different types of central places outlined by Walter Christaller can be quantified by a study of retailing outlets. SEE ALSO **central place theory**

revolution A complete circuit of the Earth around the Sun. Each revolution takes 365.25 days. Because of the tilt of the Earth's axis at 23.5°, the revolution causes the four seasons that are experienced in temperate latitudes. In spite of the tilt and the revolution, there are no major seasonal changes near the Equator.

rhyolite SEE **volcanic rock**

ria A drowned river valley in a hilly landscape. Where a river valley has been flooded, either by a rise in sea level or because the land has sunk, quite small rivers can develop large estuaries. The cross-section of a ria is V-shaped, in contrast to the cross-section of a fjord which is U shaped. Rias often have branches, which are old tributary valleys, and the estuary becomes wider and deeper toward the sea. They provide sheltered harbors, and have been used for fishing in the past, but are important for sailing in many places now. On a larger scale, when a **discordant coastline** has been flooded, large inlets are created and these are also called rias. Good examples can be seen in southwest Ireland, northwest Spain and Hong Kong.

ribbon lake SEE **finger lake**

rice A cereal, *Oryza sativa,* which is most commonly found in monsoon regions, but also grows well in equatorial climates. Rice is grown in the USA in the Gulf states and California, and also in the Po valley in Italy, the Camargue in southern France and in places along the Mediterranean in Spain. It gives a high yield per acre, and two or

three crops can be grown per year if heat and water are available. The most productive areas are on the deltas of Asiatic rivers such as the Si in southern China, and the Ganges, Irrawaddy, Mekong, Menam and Red rivers. Rice feeds more people than any other **cereal**. In the last 20 years, productivity has been increased by the development of new varieties by the International Rice Research Institute in the Philippines. The miracle rice was IR8, and many new varieties have been grown since then. Each new variety encounters difficulties with pests after a few years, and so it is a permanent research task to produce different varieties. SEE ALSO **green revolution, high-yield variety, hybridization**

Richter scale A logarithmic scale for measuring the magnitude of an **earthquake**. It was developed by American seismologist C.F. Richter during the 1930s, and has now replaced the **Rossi-Forel** and **Mercalli** scales. Earthquakes are measured on a scale of 0–9, and a measurement of 8 or over represents an earthquake of great intensity.

ridge A narrow elongated area of high ground with low ground on both sides. The sides are generally quite steep, and a ridge is likely to consist of harder rock than the neighboring rocks.

ridge of high pressure An elongated area of high pressure which is an extension from an **anticyclone**. It is similar to, though rather wider than, a **wedge** of high pressure. It gives weather similar to that of the anticyclone but is unlikely to last as long.

rift valley A valley resulting from **earth movements** along fault lines, where two or more roughly parallel faults have caused land to sink by downfaulting. Normally greater in length than width, they can be quite small or very extensive; for example, the Great Rift Valley of East Africa extends for nearly 3,000 miles [5,000 km], from Syria along the Jordan Valley, the Dead Sea, the Gulf of Aqaba, the Red Sea, and south to lakes Malawi and Tanganyika. The part of Africa east of this line is splitting away from the rest, and will gradually drift east to form a new land mass. Other rift valleys include the Rhine valley near Strasbourg, and the Great Glen between the Grampians and the Northwest Highlands of Scotland. The steep or near vertical slopes caused by the faulting are quickly worn down by erosion; even so, rift valleys are often very steep sided. ALSO CALLED **graben**

rill A small channel formed by water erosion.

rime A type of frost caused by supercooled droplets which freeze on to exposed objects. Rime drifts along on gentle breezes and freezes on to trees, telegraph poles or anything solid. It occurs on the windward side of objects, building up feathery crystals.

ring city A type of conurbation consisting of a circular urban development enclos-

RIFT VALLEY

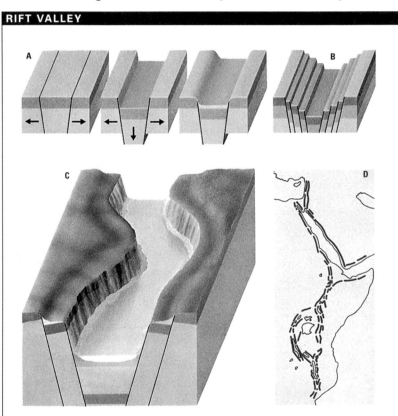

Rift valleys form when tension between two roughly parallel faults (A), causes downward earth movement resulting in the formation of a *graben* (trough of land between two faults). Sometimes a number of parallel faults result in land sinking in steps (B). A typical example of a step-faulted rift valley is shown (C).

A series of block faults can occur on either side of a graben, sometimes tilting in the process of creating the block-faulted rift valley. The East African Rift Valley (D), also known as the Great Rift Valley, is perhaps the world's best example of this type of geological formation.

ing a central **green belt**. A prime example is **Randstad** in the Netherlands.

rip Disturbed and turbulent water, usually in the sea (sometimes in a river), resulting in a strong current moving offshore. A rip may occur when two tidal streams meet, when a tidal stream flows over an irregular floor, or when a tidal stream flows into shallow water.

risk The possibility of different outcomes occurring as a result of a particular action, when the possible outcomes are known and can be calculated for.

river A channeled flow of water, which moves downhill because of gravity. Most rivers flow into the sea, although there are some which flow into lakes, and some flow into deserts, where they evaporate to leave salt flats or very saline lakes. Those which do not flow to the sea are called areas of inland drainage. Most rivers flow throughout the year, although the volume of water will vary. Some rivers, especially in deserts and regions with marked variations in rainfall, may dry up for a few months each year. The source of a river is generally in an area of high ground, and it may be a spring, a lake, a marshy patch, or a place where the **water table** reaches the surface. A river's route is called its "course." From its source, it generally flows through mountains, then through an upland area and finally on to a lowland, before entering the sea in an **estuary** or through a **delta**. The upper parts of a river are usually rocky and turbulent, and the narrow valley is V-shaped with steep sides. There may be waterfalls to interrupt the **long profile**. In the middle course the valley becomes wider, and **meanders** begin to develop. **Bluffs** and small flood plains may be forming. The lower course passes through a wide **flood plain**; the meanders are larger, and some may have been abandoned to form **oxbow lakes**.

river basin SEE **catchment area** [def. 1]

river capture The capture of a river's water supply by a second river, which works

back uphill by headward erosion and intercepts the flow. ALSO CALLED **river piracy**

river cliff SEE **bluff**

river piracy SEE **river capture**

river profile A drawing of a section of a river, whether a **long profile** or a transverse profile which is cross-sectional and can be very variable. In the hilly upper section of the river, the valley is likely to be steep, narrow and V-shaped. Further downstream, as the river grows, meandering begins, the valley widens and the profile becomes more U-shaped. Once out on the plain, the cross-section is very flattened and U-shaped. In all parts of the valley, the cross-section is likely to be asymmetrical.

river regime The pattern of seasonal variations in the volume of a river. A river regime can be shown in the form of a graph. In **monsoon** and **savanna** areas, the high points on the graph will be in summer, which is the rainy season, and some rivers may dry up altogether in the winter. The regime of a river is determined largely by the rainfall, but the local rock type will also have some influence. In

WORLD'S LONGEST RIVERS (by order of size)			
River name	Location of mouth or confluence	Length km	miles
Nile	Mediterranean	6,670	4,140
Amazon	Atlantic Ocean	6,450	4,010
Chang Jiang (Yangtze Kiang)	Pacific Ocean	6,380	3,960
Mississippi/ Missouri	Gulf of Mexico	6,020	3,740
Yenisey/ Angara	Arctic Ocean	5,550	3,445
Huang He (Hwang Ho)	Pacific Ocean	5,464	3,395
Ob-Irtysh	Arctic Ocean	5,410	3,360
Zaïre/Congo	Atlantic Ocean	4,670	2,900
Mekong	Pacific Ocean	4,500	2,795
Amur	Pacific Ocean	4,400	2,730

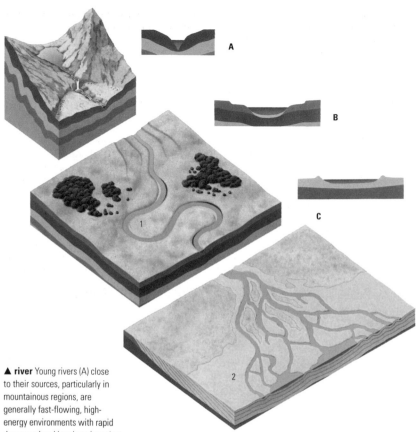

▲ **river** Young rivers (A) close to their sources, particularly in mountainous regions, are generally fast-flowing, high-energy environments with rapid downward and headward erosion, despite the hardness of the rock over which they flow. Steep-sided, "V-shaped" valleys, waterfalls, and rapids are characteristic features of young rivers. Mature rivers (B) are lower-energy systems. Erosion takes place on the outside of bends, creating looping meanders (1) in the soft alluvium of the river plain. Deposition occurs on the inside of bends and on the river bed. The process of erosion and deposition can lead to substantial changes in the course of the river. At a river's mouth (C), sediment is deposited as the velocity of the river slows. As the river becomes shallower more deposition occurs, forming an estuary or sometimes forming islands and braiding the main channel into the multiple, narrower channels of a delta. As the sediment is laid down (2), new land if created and the actual mouth of the river moves away from the source into the sea or lake.

granite areas there is always a rapid **surface runoff**, whereas in areas of chalk, limestone or porous sandstone much of the water goes underground and is released gradually during the next few days. The variations in speed of **throughflow** obviously influence the quantity of water reaching the river. When urban areas are built with a network of drains, it often means that rainwater gets into the river much faster than previously, when there were trees, soil and rocks to slow down the rate of runoff.

river terrace An old and abandoned **flood plain**. When **rejuvenation** gives a river renewed powers of erosion, it is able to cut its valley down to a lower level. When this happens, the former flood plain is left at a higher level, and a new plain is created. The old plain is called a "terrace," and in many valleys flattish areas can be seen above the level of the river. They are usually quite small. Terraces often occur in pairs, with one on each side of the river.

Roaring Forties In the southern hemisphere, the area between latitudes 40°S and 50°S, where the westerly winds blow uninterrupted past any land mass. The winds are persistent and strong, much more so than the equivalent winds in the northern hemisphere, which blow over Britain and Western Europe, and on to the northwest USA and British Columbia in Canada. In the Roaring Forties, a series of **depressions** continually moves from west

to east, bringing windy and wet conditions but quite mild temperatures.

robber economy The utilization of a resource which cannot be replaced. This is inevitable in the case of minerals, but it has also been practiced in other situations. For example, soil was used wastefully by early settlers in the western USA. Forestry can be a robber economy, unless trees are planted to replace those which have been cut down, and frequently fishing has been another example, but there are now international agreements to restrict the size of catches.

roche moutonnée *(French)* A rock which has been scraped smooth by ice on one side, but has been plucked into a jagged shape on the other, downhill side. *Roches moutonnées* can be only a few feet

RIVER CAPTURE

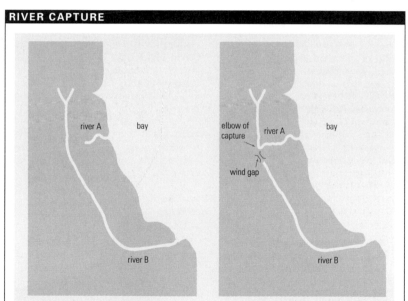

Initially, river A flows on a short course to the sea and river B a much longer course, each from the same high ground. Over time, A extends its course uphill. Eventually, the head of A meets B, and water from the upper reaches of B starts to flow through A to the sea. Thus the upper part of B becomes the upper part of A. Commonly, at the point of capture there is a sharp turn, known as the "elbow of capture." The "wind gap" is the old course of B, now dry because there is insufficient water.

in height, but some are much bigger, reaching 100 ft to 130 ft [30 m to 40 m]. They are found in glacial valleys; when viewed from up the valley they will appear smooth and gentle, but viewed from downstream the jagged rocky edge will be visible, and the landscape will look very different. The smooth surface will have scratch marks, or **striations**, caused by ice and small rocks scraping over it. The jagged downstream side will have suffered from **freeze-thaw** activity. It is said that *roches moutonnées* were so named because they look rather like sheep lying down, or perhaps like old-fashioned wigs called *moutonnées*. A good imagination is required to see the resemblance!

rock A collection of minerals which have been cemented together. Rocks make up the surface of the Earth, and they are very varied in appearance, hardness and mineral composition. All rocks can be classified into one of three groups: **sedimentary**, **igneous** or **metamorphic**. Sedimentaries consist of sediment, generally carried by rivers down to the sea where it all accumulates, that is then compressed and

▼ **roche moutonée** Formed by glacial activity, and thus found in glacial valleys, a roche moutonée looks quite different from the top (the left in the diagram) and from the bottom (the right) of the valley. These features vary considerably in size.

uplifted to form new rocks. They are composed of land sediment together with fragments of dead sea creatures. Igneous rocks are those which result from volcanic activity, either underground (intrusive) or out on the surface (extrusive). Metamorphic rocks were either igneous or sedimentary, but they have been changed by the effects of heat or pressure. The heat comes from igneous activity, and the pressure is the result of **folding**. The most intensive metamorphism occurs when heat and pressure are experienced in the same locations.

rock flour Fine particles produced by **abrasion** beneath a glacier or ice sheet. As the ice moves downstream, the rocks that it is carrying scrape and grind the **bedrock**, and tiny fragments of rock flour are produced. Much of the rock flour is transported by subglacial streams and contributes to the load of glacial rivers, which are often inky in color as a result.

roll-on-roll-off A ferry or ocean ship service which transports vehicles. Trucks, buses, automobiles and other vehicles can drive directly into the ship at one end, and off at the other on completion of the crossing.

root crop A plant which is grown for its edible roots. Potatoes are particularly important for human consumption, and

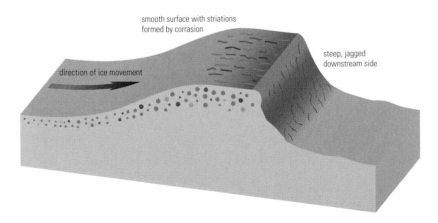

smooth surface with striations formed by corrasion

steep, jagged downstream side

direction of ice movement

sugar beet is an important source of sugar. In tropical lands cassava, yams and sweet potatoes are vital to many groups of people. Turnips, rutabaga and several other root crops are a common source of cattle feed.

ropy lava An extrusion of lava which has solidified in extended linear fashion, like pieces of thick rope. A thin skin forms on the lava while it is still moving, but the molten liquid beneath the skin continues to move for a time to produce an elongated shape. ALSO CALLED **pahoehoe**

Rossi-Forel scale A scale for measuring the intensity of **earthquakes**, devised in 1878 by M.S. de Rossi and F.A. Forel. It was commonly used until 1931, when the modified **Mercalli scale** became more popular. Both have now been superseded by the **Richter scale**.

Rostow's model A model describing the stages of economic development which all countries will pass through. It was produced in 1955 by W. Rostow, who believed that all countries will follow a similar pattern of development as they progress from a traditional agricultural economy. There are five stages to his model:

Stage 1. The traditional society is still largely agricultural, often at a subsistence level, and there is little utilization of resources. Some areas of the world are still at this level.

Stage 2. Preconditions for take-off. Some modern methods of agriculture may be introduced, and outside influences such as colonialism will bring about changes. There will also be the beginnings of exploitation of resources.

Stage 3. Take-off. This means the beginning of industrial development, which may be initiated by the use of resources, such as coal in the early 19th century, or the exploitation of oil in the Middle East in the 1960s. Earnings increase, and there is money available for investment. Agriculture becomes more commercialized.

Stage 4. The wealth created in Stage 3 provides the basis for expansion of industry. This is called the drive to maturity, and

growth occurs in all sectors of the economy. Money is available to buy luxuries and consumer goods, which provides a boost to many other industries. Technology and technical skills increase.

Stage 5. The stage of high mass consumption. Urbanization has resulted in well over 50% of the total population living in cities, and this figure reaches as much as 90% in some countries. The affluent society, as in the USA, Western Europe and Australasia, spends more on consumer goods and services. The major area of employment is in the **tertiary sector**, with an increasing number of **quaternary sector** jobs, too.

The older developed countries have passed through the stages at a fairly slow rate, but many of the **Third World** countries are progressing through the five stages within a few decades. There may be different regions of the same country experiencing each of the five stages at the same time; for example, Brazil, Mexico and India.

rotational slip The downward movement, in a semicircular motion, of a mass of rock, soil or ice along a concave face.

rotation of the Earth The turning of the Earth about its own axis. One complete rotation takes 24 hours, during which day and night are experienced, as different parts of the Earth turn to face the Sun. Because of the tilt of the Earth's axis at 2.5°, the length of day and night varies. Without the tilt, they would be of equal length. At the Equator day and night are approximately of equal length throughout the year. At the North Pole there is 4.5 months' daylight and 4.5 months' darkness, while for the rest of the year there is twilight. The same is true of the South Pole, although the periods of day and night are reversed. All places north of 66.5°N and south of 66.5°S receive at least one 24-hour period of continuous daylight. However, in a complete year, all parts of the Earth receive exactly the same total amount of daylight, although the daily and monthly distribution is variable. On the **equinoxes** – 21 March and 22

September – everywhere has 12 hours' darkness and 12 hours' daylight.

rough grazing Grassland that is low in quality and unimproved. Some moorland, scrubland, marshland and mountain pasture is described as rough grazing. It is often unfenced and used for extensive grazing of sheep or, occasionally, beef cattle. Rough grazing can be improved by the addition of fertilizers and better, sown grasses. Some parts of the world do not distinguish between different qualities of pasture, and merely divide farmland into **arable** and **pastoral**.

rubber An elastic material, obtained from the **latex** of various species of *Hevea* and *Ficus*. The most important rubber tree is *Hevea brasiliensis,* which originated in the Amazon basin of South America, but is now grown on a larger scale in plantations in Malaysia, Indonesia, Sri Lanka and elsewhere. The native rubber trees are still used in Brazil, but production is very small and few plantations have been developed. The town of Manaus in the middle of the Amazon basin remains a small market, but a century ago it was the rubber center of the world. Synthetic rubber can now be produced in chemical works, using petroleum as the raw material. The USA and Germany manufacture large quantities of synthetic rubber, and so the demand for natural rubber has declined.

rudaceous rock A coarse-grained **sedimentary rock**. Rocks which are derived from land sediment are split into three main groups according to grain size. If the particles are coarse, like large grains of sand or gravel, the resulting rock when consolidated is rudaceous. Examples include **breccia**, **conglomerate**, **gravel**, **scree** and some types of stony boulder clay. Less coarse-grained rocks are classed as **arenaceous,** which means sandy, and fine-grained sedimentary rocks are known as **argillaceous** rocks, or clays.

runoff SEE **delayed runoff, surface runoff**

rural depopulation The movement of people from rural areas into the cities. Such movement has been going on for centuries, but has quickened in more recent times. People are attracted to the cities by the prospect of bright lights, more jobs and better pay. The movement grew rapidly in many countries after the **Industrial Revolution**, and urban areas expanded dramatically. As more jobs and more money became available, rural inhabitants left the land, so that many farms changed to more pastoral activities, which required less labor but were often equally profitable to the landowners. As the industrial developments grew, machines were provided to lighten the burden of labor on the farms, and this enabled more people to leave and drift into the cities. In more recent times, especially since 1950, there have been large-scale movements from rural areas into the cities in **Third World** countries. Unfortunately, there are often not enough jobs available, and so many of the migrants are very poor and have to live in **shanty towns**. SEE ALSO **migration**

rural settlement A settlement in a rural area, typically a small unincorporated village or town. Formerly, rural settlements were associated with working in agriculture, but in North America and Western Europe this is not always true, as only a very small proportion of the working population is employed on farms. Many rural settlements are now occupied by city dwellers. The distribution of rural settlements in modern and industrialized environments cannot really be related to the **central place theory** of Walter Christaller.

rural-urban fringe The transitional area between town and country. It is not always clear where an urban area ends, and even beyond its limits it will still be very influential on the people living in the surrounding rural areas. The idea of a rural-urban continuum is found in North American countries, Europe and Japan. There is now a graduation from urban to rural areas, as urban influences stretch out into the country, and many rural dwellers have their own transportation, enabling them to use the town's facilities. The effect of distance has decreased because of the increase in mobility. Rural areas depend on towns and cities for jobs, supplies of food, stores, hospitals, schools, and also for administration, which is based in the larger cities. In **Third World** countries, many people are still involved in agriculture, and many do not have transportation and mobility.

S

saddle A slightly lower area between two summits. It is often flattish, and provides a **gap** for a route between two valleys. It is similar to a **col**, but is probably slightly broader.

saeter An area of mountain pasture. Located above the forests of the mountain side, it will be used for summer grazing. Most families from the villages in the valleys have a hut on a saeter, and one or two members of the family move up to the hut to look after the cattle while they are on the pasture. The cattle are milked, and butter or cheese is made, although some milk may be transported down to the valley. This use of mountain pasture is very similar to alpine farming in Switzerland, France and Austria. ALSO CALLED **seter** *(Norwegian)*

sagebrush A type of thorny **xerophytic** scrub vegetation, which can survive in semiarid locations. The southwest USA has large areas of sage brush, and Nevada is known as the "Sagebrush State."

Sahel I A large area of Africa to the south of the Sahara Desert. It stretches from Senegal and Mauritania in the west to Sudan and Ethiopia in the east, and includes parts of Burkina Faso, Nigeria, and many other countries. It has been notable for famines in recent years, and the subsistence farmers living in this area have found life very difficult. Overpopulation, as well as too many animals, has prevented the resources of the area from supporting all the people. Overgrazing has ruined the vegetation, and the land has been exposed to **soil erosion**. The bare land has lost its soil so no vegetation can grow. The Sahara Desert has been expanding southward into the new man-made desert. **Desertification** has been accelerated by the fact that most of the local people use firewood for fuel, and if any trees can be found they are used as fuel, so that there is nothing left in the ground to protect the soil. Heavy rain washes away the bare soil, and wind can blow away the fine particles. **2** The semiarid lands in Tunisia, along the margin of the Sahara Desert. Although not a wealthy area, this region does not have the same kind of problems as the land to the south of the Sahara.

saki A means of lifting water from one level to another. It is rather like a seesaw with a bucket at one end, and is an important method of irrigation in parts of the Middle East and the Nile valley in North Africa. ALSO CALLED **sakiyeh**

sakiyeh SEE **saki**

salina *(Spanish)* A saltpan. Salinas often form in desert or semiarid regions where a stream has flowed into an area of inland drainage, and then evaporated, leaving behind tiny deposits of salt. The deposits can accumulate for hundreds of years to form large salty areas, examples of which are found in parts of the southwest USA, in the interior of Australia, and on the Andean plateaus of Bolivia.

salinity The degree of saltiness, as of the oceans. The average proportion of salt in the sea is about 35 parts per 1,000, or 3.5%. In the Red Sea, because of high evaporation rates, the figure is over 40, but in the Arctic it falls to less than 30. In the northern parts of the Baltic Sea it is less than 10. In the Great Salt Lake in Utah, USA, the figure is about 200 and in the Dead Sea it is about 250. The salts found in the oceans include sodium chloride (common salt), which accounts for about 75%, magnesium chloride, magnesium sulfate and calcium sulfate.

salt Sodium chloride ($NaCl$). It is found in sheets and layers around salt lakes, as well as in old strata, where it is a relic of desert conditions of millions of years ago. Salt often folds up to form domes, which

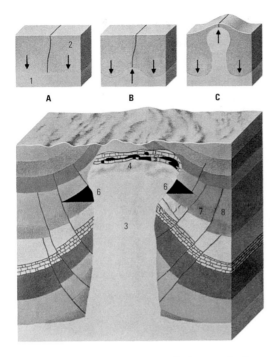

◀ **salt dome** A salt dome, or diapis, is caused by a massive plug of salt that has been forced upwards through a fault in overlying sedimentary strata by subterranean pressure. Oil and gas are often trapped in rock strata associated with salt domes.

KEY
1 halite beds (salt)
2 sedimentary overburden
3 salt plug
4 gypsum and anhydrite
5 porous limestone caprock
6 oil trap
7 sand
8 shale
A sedimentary beds overlying halite beds
B pressure of denser overlying beds causing peripheral sinking and salt to rise up fissure
C formation of salt plug and dome

are not only important sources of salt, as in Texas and neighboring states in the USA, but also make good reservoirs for oil and natural gas.

saltation A process whereby sand grains moved by the wind bump into other grains, which are then moved forward as well. In this manner, sand is moved downwind in desert areas, as well as in areas of dunes. The same term is used for the movement of sand along the bed of a river, in which the flowing current dislodges grains of sand and pebbles.

salt dome A body of salt that has intruded into a sedimentary rock overlay. The flow of the relatively plastic salt into a dome may be the result of a difference in density between the salt and the overlying rock, but the process is not completely understood. ALSO CALLED **diapsis**

salt lake A saline lake located in a desert area, probably in a region of inland drainage.

As the water evaporates the salinity is likely to increase, and around the shores of the lake, layers of salt will solidify. Examples can be seen in all desert areas, but among the largest are the Great Salt Lake of Utah, in the USA, and Lake Eyre in Australia. If there is a spell of exceptionally wet weather, much of the salt may be dissolved and washed away.

salt marsh An area of coastal marsh which is flooded by the sea from time to time. Salt marshes are also found around **salt lakes** in deserts, and near large lakes and estuaries in warm regions, such as the Camargue in southern France or the Guadalquivir delta in southern Spain. Special salt-resistant plants colonize the areas; they trap silt and help to build up the land. Many of the coastal marshes can be reclaimed once the land has become well vegetated.

sample A small part of a full population taken to represent the characteristics of the whole of that population. There are three main types of sample: systematic sample, where survey points are chosen at equally spaced intervals; random sample, where points are chosen at random intervals; and stratified sample, where the area being studied is divided up into different parts and random samples are taken from each group.

San Andreas fault A fault, 700 miles [1,125 km] long, stretching through California in the USA. It marks the boundary where two crustal plates move alongside each other, with the friction causing **earthquakes** on the surface. There are mountains on either side of it, caused by the movement of the plates. The cities of San Francisco and Los Angeles lie near the fault and have suffered many earthquakes and subsequent damage; it is widely predicted that a large earthquake is due to happen in this area within the near future – one in 1906 killed 3,000 people and nearly destroyed San Francisco.

sand dune A ridge of sand formed in deserts and along coastlines wherever there is loose material which is not anchored by vegetation. Sand particles are transported by the wind and gradually move in a downwind direction. Sand dunes form along sandy shorelines, and dry sand is gradually blown inland. It piles up into a series of ridges, with lower hollows or **slacks** in between. Marram grass often colonizes the dunes and helps to anchor them. In deserts there are often large expanses of sand sheet, but there are two distinctive dune shapes; these are **seifs** and **barchans**.

sandstone A **sedimentary rock**, in which the grains of sand are 0.001 in to 0.1 in [0.02 mm to 2 mm] in diameter. If the grains are larger, the rock is a **rudaceous** rock or gravel, and if smaller the rock is silt or clay. Sandstones are **arenaceous** rocks, and are very variable in character as a result of differences in grain size, chemical content and cementing materials. Most sand grains contain **quartz**, which is hard and resistant. Other minerals in sandstones

include **feldspars** and **micas**, and iron also occurs, which tends to give sandstones a reddish or brownish color. Most sandstones are **porous** and produce poor soils for agriculture, though they can be improved quite easily with the addition of fertilizer. They give light soils suitable for vegetables, and other crops. Some sandstones are very hard and give rise to high ground. The majority of sandstones, though, are between the two extremes, and form undulating countryside. Most sandstones are formed by the accumulation of river sediments on the seabed. They are then compressed and uplifted to form new land. There are also a few sandstones which have been formed of windblown materials, especially in desert regions.

sandur SEE **outwash plain**

saprophyte A plant which lives on decaying vegetable matter. Fungi are good examples of this type of plant. Saprophytes are vital decomposers in ecosystems.

satellite photographs There are numerous satellites circling the Earth today. Some are in geostationary orbit, which means that they remain over the same point of the Earth, traveling at the same speed as the rotation of the Earth. Others circle around the Earth in a polar direction, and with each orbit pass over the Equator about 30° west of the previous orbit; this is because the Earth continues to rotate below the satellites. The photographs have daily use for **weather** forecasting, but can also be used in many other ways. For example, they may provide information for mapping land use, plotting water temperatures in the oceans, or marking the position of icebergs. Satellite photographs are images recorded electronically rather than traditional photographs. LANDSAT satellites send down information on land use, and METEOSAT gives information on cloud cover.

satellite town A self-contained town found well away from a major town, but still associated with it, especially for communications. For example, many of the **new towns** in Britain were satellites of London.

saturated In **meteorology**, applied to air that cannot hold any more water vapor. If cooling of the saturated air occurs, there will be condensation.

saturated adiabatic lapse rate The rate at which saturated air cools as it ascends. The average rate is 1°F per 350 ft [1°C per 200 m], but it shows considerable variation, depending on the air temperature and the amount of water vapor present. The rate is much lower than in dry air, because the moisture gives off latent heat when it condenses. SEE ALSO **dry adiabatic lapse rate**

savanna An area of tropical grassland, found between the major desert regions and the tropical forests. Savannas occur mainly between 5° and 20° north and south of the Equator, although, in East Africa, savanna regions are situated on the Equator as well. This is because the land is elevated, and the East African plateau is at the appropriate height for grassland vegetation to grow. Both the rainfall and temperatures are too low for tropical forest. The savanna vegetation is supported by the savanna type of climate. This is hot and wet in summer, when the sun is overhead; temperatures average about 80°F [25°C] and the rainfall is 10 in to 50 in [250 mm to 1,250 mm]. In winter the tropical high-pressure system controls the climate, and so the conditions are warm but very dry. Temperatures are around 60°F to 75°F [15°C to 25°C], and rainfall totals are less than 10 in [250 mm]. In such a climate, grass grows very well in summer – up to 6 ft [2 m] in height – but it will shrivel in winter. Trees do not grow well, although a few acacias and eucalyptus are able to survive the long dry winters, and baobabs and bottle trees survive by storing water in their trunks. Trees are most numerous at the edge of the savanna which is nearer to the tropical forest. In the northern hemisphere, this is the southern edge; for example, in Nigeria and Venezuela. It is the northern edge for the southern hemisphere savannas, such as the Brazilian plateau and parts of Australia. **Pastoral farming** is the main activity on the grasslands, and this may be large-scale commercialized farming, as in Northern Territory of Australia, or subsistence and seminomadic, as for the Masai or **Fulani** in Africa. Coffee is grown on plantations in Brazil, and tobacco in Zimbabwe, but corn is generally the main food crop grown. There are many wildlife parks in African savanna areas, where giraffes, zebras, lions and other animals can be seen. The savannas are sometimes called "sudan" areas in eastern Africa, and are known as *campos* in Brazil and **llanos** in Venezuela.

scale The size of a map expressed as a ratio of the actual size of the land. SEE ALSO **representative fraction**

scar A short steep slope, which is generally bare rock. The term is used in northern England, and it is generally associated with areas of **Carboniferous** limestone.

scarp SEE **cuesta**

scatter diagram SEE **scatter graph**

scatter graph A graph which shows the relationship between two variables. A line of best fit can be drawn on the graph to show the average trend, and then the extreme examples, above and below the average, will be clearly shown. One of many ways in which a scatter graph could be used would be to put the population of settlements along the horizontal axis, and their height above sea level on the vertical, to determine if there were a relationship between height and size. ALSO CALLED **scatter diagram**

schist A **metamorphic rock** characterized by thin plates or flakes of minerals, overlapping rather like the tiles on a roof. Almost any type of rock will become a schist if subjected to sufficient metamorphism, and it will be named after its most common mineral; for example, mica schist.

SCATTER GRAPH

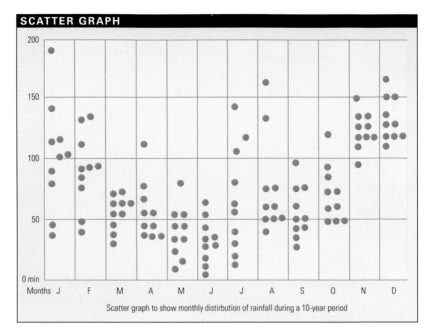

Scatter graph to show monthly distirbution of rainfall during a 10-year period

schistosomiasis SEE **bilharziasis**

scree Loose fragments of rock which have accumulated on a hillside. They are likely to slide down the hill when lubricated. The rock fragments may be very varied in size, and a small amount of sorting will take place as the particles move downhill, with the larger particles tending to move to the bottom. Scree forms more frequently on certain rock types, such as limestone. **Freeze-thaw** activity and the effect of changes in temperature are the most important processes involved.

scrub A type of vegetation found in locations which do not have rainfall throughout the seasons. Low shrubs and small trees up to 6 ft [2 m] in height are common, and the growth may be quite dense. Good examples are found in a variety of locations. There is the **maquis** in the Mediterranean lands, the **mallee** in Australia, and the **chaparral** in California. There is also much scrub along the margins of deserts, between the **savanna** and the more arid lands; for example in West Africa. There is much scrub

in Brazil and also in neighboring Paraguay, where the Chaco region is a good example.

sea breeze A wind which blows from the sea on to the land. In tropical areas it is usually a cooling breeze, as the sea is cooler than the land. In **temperate** latitudes, a wind from the sea can feel cool and raw, and can be bitterly cold if coming from the Arctic or any icy sea. A daily sea breeze often develops in settled anticyclonic weather. As the land heats up during the day, air begins to rise. The land becomes warmer than the sea, and, as the air pressure over the land is slightly lower, a wind will begin to blow from the sea. SEE ALSO **land breeze**

sea-floor spreading SEE **ocean-floor spreading**

seasons One of the four periods of the year: spring, summer, fall (or autumn), and winter, which are characterized by different climatic conditions. In Arctic regions, spring and fall are very short, and summers are fairly cool. Near the Equator there are virtually no seasonal differences. In the southern hemisphere, however, the situation is reversed, with winter during June and July, and summer in December and January.

sea surge A large movement of oceanic water, which is likely to cause large waves or high tides when it reaches land.

Secondary A geological era, now generally known as the **Mesozoic**.

secondary depression A small area of low pressure which is found on the edge of a major depression. It may just be a bulge in the shape of the **isobars**, or it may be a

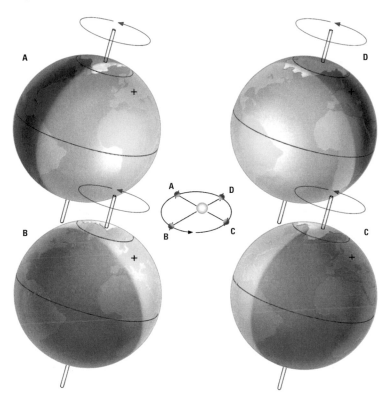

▲ **seasons** Each of the four seasons begins and ends at one of the two solstices (the shortest and longest days of the year), which are separated by six months, or one of the equinoxes (when day and night are equal), also separated by six months. The solstices mark the days when one pole is at its closest position to the Sun and the other is farthest away from it. Equinoxes mark the days when each pole is the same distance from the Sun. For example, spring in the Northern Hemisphere at the spot marked by a cross on the globe commences on the equinox of March 20 (A), when day and night are equal length, and is followed by the summer solstice on June 20 or 21 (B) (the dates vary because of leap years),

when the North Pole is at its closest to the Sun and the day is the longest of the year. The fall equinox commences on September 22 or 23 (C), when day and night are once again equal in length. Winter in the Northern Hemisphere beings on December 21 or 22 (D), when the day is shortest in the year. This seasonal sequence is reversed in the Southern Hemisphere, where, for example, winter starts on June 20 or 21 when the South Pole is farthest away from the Sun but the North Pole is closest and the Northern Hemisphere's summer begins. The start and finish of each season can thus be defined in very precise astronomical terms by the Earth's position relative to the Sun.

separate system with its own set of roughly circular isobars. Secondary depressions will often develop to become deeper and more intense than the primary depression. Like all depressions, they gradually fill in and become weaker. SEE ALSO **depression**

secondary industry An industry which processes materials produced by the **primary sector**. Secondary industry is considered to include the building industry.

secondary sector The part of the economy which deals with manufacturing and processing; for example, making steel, automobiles or clothing. Employment in secondary activities increases as countries become industrialized, but decreases later as the **tertiary sector** develops. The tertiary sector provides services; the **primary sector** produces raw materials; and the **quaternary sector** provides information.

secondary vegetation Vegetation that grows in an area once cleared of its primary vegetation; for example, after a fire or settlement by man.

second home A home used only occasionally by its owner, whose usual residence is elsewhere. Such homes are often found in rural areas or by the sea, as they are commonly used for recreational purposes at the weekends or during holidays. Local residents of areas where there are many second homes may be forced out as house prices are driven up.

Second World Those countries in the developed world with a centrally planned communist economic system, in particular the USSR and Eastern Europe before the political changes in 1990–1. SEE ALSO **First World**, **Third World**

sector model One of the three classical models of **urban morphology**, described by **Hoyt** in 1939. He used sectors as the basis of the divisions of land use in a city, and believed that transportation routes would be a major influence on the growth of urban

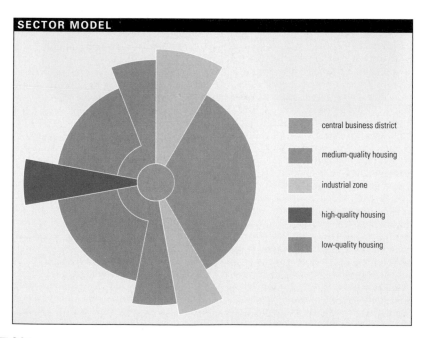

SECTOR MODEL

- central business district
- medium-quality housing
- industrial zone
- high-quality housing
- low-quality housing

areas. He described his model of urban structure after analyzing land-use patterns in 142 American cities. His five main types of land use were: 1 the **central business district (CBD)**; 2 industrial zones; 3 low-quality housing; 4 medium-quality housing; 5 high-quality housing. He added his ideas to those put forward by Burgess in his concentric model, but it is still an oversimplification of reality. SEE ALSO **Burgess model**, **Harris and Ullman model**

sedentary agriculture Settled agriculture in a fixed location, as opposed to nomadic or shifting agriculture. All commercial crop growing has to be sedentary. With the exception of some **transhumance** movements, **pastoral farming** is also sedentary, unless it is for subsistence only.

sediment Rock fragments which have been formed by **weathering** and **erosion**. They can be very varied in size. Once broken down, they are moved by water, and in certain places by ice or wind, and will be transported largely into low-lying areas. Most sediment is eventually deposited in the sea.

sedimentary rock Any rock formed by deposition of **sediment** derived from preexisting rocks, which may have been sedimentary, **igneous** or **metamorphic**. Most sediment accumulates on the bed of the sea, having been dumped there by rivers, or accumulated there as dead sea creatures fell to the ocean floor. The accumulated sediment will be consolidated and compressed. **Earth movements** uplift the sediments, and they may be tilted, folded or faulted. The resulting rocks are sedimentary, and their type depends on their composition. The layers in sedimentary rocks are called strata; they may be a few inches or many feet in thickness. Sedimentaries consisting of land sediment are mechanically formed or "clastic" rocks, and are gravels, sands, silts or clays, according to the size of the particles. Other types of sedimentary rock include limestones, which consist of fragments of dead sea creatures, sometimes mixed with land sediment; evaporites, which are saline deposits; coal, which is accumulated vegetation; coralline, which contains large quantities of coral; and chalk, which is a pure form of limestone, with very little land sediment.

seeding of clouds The practice of putting more nuclei into clouds in order that more water droplets will condense on to the nuclei and form sufficiently heavy droplets to fall as rain. Dry ice and silver iodide are used as nuclei, and experiments have been conducted in the USA, Australia, southern Africa and the Middle East with the aim of producing rain in areas suffering from a shortage. In some cases, rain has fallen after cloud seeding, but it is not certain whether this was as a consequence, or just coincidence.

seif An elongated type of sand dune, found in parts of Libya and other areas of sand desert. Often occurring in groups, seifs may extend for several miles. The narrow ridge will run parallel to the prevailing wind direction; when cross-winds blow, the height and width of the seif can be increased.

seismic focus SEE **earthquake**

seismograph An instrument which measures **earthquakes**. The vibrations are recorded by a pen on a revolving drum. Some seismographs are so sensitive they can pick up seismic activity thousands of miles away. For example, an earthquake in Los Angeles in California, USA, in January 1994 was picked up by seismographs in Europe.

seismology The study of earthquakes.

self-sufficiency 1 Subsistence farming, in which farmers grow enough food for their own requirements, and also make their own clothes. 2 The ability, especially of a developing country, to meet its own requirements for industrial goods.

selvas (*Portuguese*) A dense tropical forest found in the Amazon basin in South

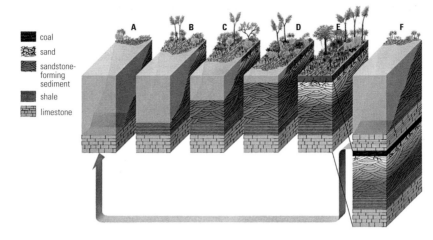

- coal
- sand
- sandstone-forming sediment
- shale
- limestone

▲ **sediment** A delta's sediments are laid down in a specific order that may be repeated many times over if the region where deposition takes place is sinking. Limestone deposits cover the seabed when the delta is too distant from a given point to be influential (A). As the delta encroaches (B), fine-grained muds that will later become shale are deposited, followed by coarser, sandstone-forming sediments as the advance continues (C). As the water becomes shallower, current bedding (D) indicates that sand is being deposited. Once the delta builds above water level (E) it can support swamp vegetation, which will eventually form coal. When the region sinks (F) the cycle restarts.

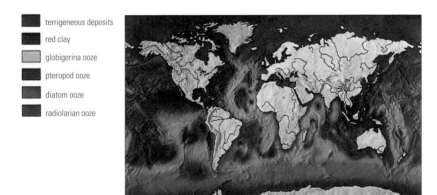

- terrigeneous deposits
- red clay
- globigerina ooze
- pteropod ooze
- diatom ooze
- radiolarian ooze

▲ **sediment** The seafloor around the world is covered by unconsolidated sediments that are classified according to the nature of their main constituent. This constituent is determined by the distance from the landmasses, the nature of the winds and currents, the surface water temperature and the depth. The major types are: the terrigeneous deposits, which are debris derived from the weathering of the continents; the red (or brown) clays that form the inorganic sediments; and the globigerina, radiolaria and diatomaceous oozes, which are formed by the accumulation of the shells of dead planktonic animals.

America. There are similar forests in parts of central Africa and in Southeast Asia. They are characterized by a rich variety of plants and insects, many of which are not found elsewhere. Large areas of the forests are being cut down in a wasteful and extravagant way at present.

séracs A very irregular surface of ice, generally found at the foot of an **icefall** on a glacier. It is the result of a crevassed section of a glacier becoming compressed when the gradient decreases, or the speed of ice movement is reduced. Pillars and pinnacles of ice stand up several feet above the general level of the ice, and make an almost impenetrable barrier.

sericulture The rearing of silkworms for the production of silk. It is necessary to grow mulberry trees, as the silkworm lives on mulberry leaves.

serra (*Portuguese*) A range of mountains.

service Any of the tertiary industries, which help or serve the general public, including buses and trains, stores, banking and insurance. Large cities have a larger number of services than can be found in smaller settlements.

seter SEE **saeter**

settlement Any collection of dwellings, ranging from a small hamlet with two or three families to cities the size of New York or London. In rural areas there is often a small settlement every few miles, but larger urban areas are more unevenly distributed. There is generally a reason for the original choice of site, even though the reason may no longer apply, or the settlement may have grown too big for its original site. For example, water supply commonly dictates the choice of site for a settlement, but if a large town develops, the water supply will probably become insufficient. For this reason, most large cities, such as Las Vegas or Los Angeles in the USA, have to obtain much of their water from many miles away.

settlement hierarchy A system for classifying settlements according to their size and influence, as propounded by Walter Christaller in his **central place theory**.

shade temperature A temperature reading taken in the shade, usually in a **Stevenson screen**. Official temperature readings are always shade temperatures. They exclude the effects of sunshine and shade differences, and the effects of wind. The reason for this is to standardize methods, and to enable fair comparisons to be made. By using a whirling **hygrometer**, it is possible to overcome some of the effects of sunshine and wind, if temperature readings outside a Stevenson screen are required.

shading map SEE **chloropleth map**

shaduf A simple lifting device widely used for irrigation in the Middle East and elsewhere. It consists of a long suspended pole with a weight at one end and a bucket at the other. Manpower is used to dip the bucket into a well and lift it out again. Once out of the well, it is swung around and emptied into a ditch or trough. It is similar to a **saki**, but is used for obtaining water from wells, rather than from rivers.

shaft mining A mining method which uses a vertical shaft to reach and extract the mineral. This is likely to be more expensive than **adit** or **open-pit mining**.

shale A type of sedimentary rock which is similar to clay. It is fine grained and consists of thin layers or sheets. Each layer probably represents a period of **deposition**, after which there was a slight pause or change in sedimentation. The layers have been consolidated, but shales can be split easily into horizontal pieces.

shanty town An area of unplanned and spontaneous development in an urban area. Shanty towns are characteristic of large cities in the **Third World**, where thousands of migrants drift in from the rural areas in search of employment, which is generally not available. Many of the people have no money and nowhere to live. They build simple and often very primitive houses out of cardboard, beaten-out and flattened gas cans, or any other materials they can find. The shacks are often constructed at the edge of the built-up area, or on any spare piece of land which can be found in the city; for example, old quarries have been used in Rio de Janeiro in Brazil. Derelict building sites, steep slopes thought to be unfit for proper housing, land near sewage outlets, and old rubbish dumps have all been used. In Calcutta, India, there have even been squatters in a pile of large drainage pipes which were lying around. Shanties have high densities of population, which create problems of hygiene as well as crime. There are often no water supplies and no electricity. However, in many cases all over the world, shanties have been improved by the efforts of the people living in them. Stronger and more permanent housing is built by people who have managed to get a job and earn some money. In many cases, a water supply is provided, probably just one pipe and one tap for one or two rows of houses. The gradual improvement of living conditions has meant that some shanties have been turned into real suburbs. Once the occupants are given legal status and official authority to reside on the land, they work very hard to raise their standards. Most Third World countries still have shanties, and possibly as many as one-third of the urban inhabitants of these countries are classified as shanty dwellers.

sharecropping A system of farming in which the landowner is given a percentage of the crop harvested instead of a rent in cash. The share will vary from place to place, but is often as much as 50%. In some cases, the landlord, who is likely to be wealthy, may provide machinery and seeds for the sharecropper to use. Sharecropping was used in the southern USA when slavery was abolished, but is also found in many other countries, including France.

sheep track A very narrow terrace in an area of softish rocks, such as glacial drift,

where **solifluxion** has been taking place. In some of the hilly area where sheep are kept they are used by sheep as walkways and can be worn to look like narrow footpaths. Although they are referred to as sheep tracks, their formation was quite unrelated to the sheep. ALSO CALLED **terracette**

sheet erosion A form of erosion in which the surface materials are all removed. It may be the result of wind blowing away all the fine particles, or rainwash carrying away the particles after sheet flooding. Sheet erosion occurs only if there is no vegetation cover. When it occurs, it removes the fine particles of soil, leaving the coarse, less fertile parts behind. Once the land has been denuded by sheet erosion, it is likely that future rainstorms will start to cause **gully** erosion as well. Sheet erosion has been most serious in areas where all the vegetation has been removed by plowing land which was too dry, or by overgrazing. The **dust bowl** and parts of the **Sahel** have suffered particularly from sheet erosion. ALSO CALLED **sheet wash**

sheet wash SEE **sheet erosion**

shield An area of old hard rock, such as the Canadian Shield or the Baltic Shield. The Canadian Shield is located between Hudson Bay and the Great Lakes in North America. The Baltic Shield is found in southern Finland, and in parts of Sweden and western Russia.

shield volcano A volcano which has been built up by eruptions of fluid **basic lava**. Because the lava has been free flowing, the slopes of the volcano are quite gentle, allowing the lava to flow out over a large area. Eventually, a massive mountain can be created. The largest example is Mauna Loa in Hawaii, which has risen over 30,000 ft [9,000 m] from the seabed. It is higher than Mt. Everest (at 29,029 ft [8,848 m]), but half of its height is below sea level. Its base covers an area of nearly 20,000 sq miles [50,000 sq km]. The name "shield" is derived from the fact that basic volcanoes are shaped like an upturned shield or a saucer.

shifting cultivation A style of farming found in tropical forest areas in South America, central Africa and parts of Southeast Asia. The people live in a village community, and they make clearings in the surrounding forest, in which to grow crops such as cassava, yams, sweet potatoes, corn and beans. Once an area of forest has been cleared – usually by burning – it will be used for growing crops for two or three years, by which time the soil will be losing its fertility. Then the clearing will be abandoned, and the forest can regrow, giving the soil a chance to recover. Meanwhile, a new clearing will be created elsewhere in the forest. The system works well if population densities are low, but if the land has to be reused too quickly – say, in less than 30 or 40 years – the soil will not have recovered sufficiently for crops to grow successfully. ALSO CALLED **bush fallow**, **chitamene** (*in Zambia*)

shingle Pebbles which accumulate on a shore. The stones have been smoothed and rounded by water action and by **attrition**. Shingle sometimes builds up into ridges.

ship canal A canal which can be used by oceangoing vessels. The Panama Canal and Suez Canal are outstanding examples. In North America, there are ship canals on the St. Lawrence Seaway, the Welland Canal near Niagara, and the Soo Canals between lakes Huron and Superior. There is the Manchester Ship Canal in England, and in the Netherlands, Amsterdam is linked to the North Sea by the North Sea Canal.

sial The lighter, granitic layer of the Earth's crust consisting largely of silica and aluminum. The layer of sial rocks forms the continents, which sit on top of the sima layer. The sial is less dense than the sima, and is generally not carried beneath the surface when subduction occurs. SEE ALSO **sima**

sierra (*Spanish*) A range of mountains.

silage A fodder crop which has been harvested while green. It is generally stored in a silo, and provides a very nutritious winter

feed for animals. Drying and haymaking are eliminated by the use of green fodder, which is particularly helpful in wet climates.

silicon An element which comprises about 28% of the Earth's crust. It is the second most common element in the crust, after oxygen, with which it combines to form silicon dioxide (SiO_2), generally known as silica or **quartz**. It is also found in **feldspars**, **micas** and many other minerals.

sill 1 The lip or threshold of a **cirque**. It often acts as a dam, so that after the ice has melted a cirque lake is formed. 2 An **igneous** intrusion. It is a thin sheet which is almost horizontal and is concordant to the preexisting structure, which means that it has squeezed in between the **bedding planes**, and then cooled and solidified. Sills can be a few inches or many feet in thickness.

silt A sedimentary deposit laid down in a river or lake. It consists of particles which are finer than sand but coarser than clay, with a diameter ranging from 0.0001 in to 0.001 in [0.002 mm to 0.02 mm]. Consolidated silt forms siltstone, one of many **argillaceous** rocks.

Silurian Third oldest period of the Paleozoic era, lasting from 430 to 395 million years ago. Marine invertebrates resembled those of Ordovician times, and fragmentary remains show that jawless fishes (agnathans) began to evolve. The earliest land plants (psilopsids) and first land animals (archaic mites and millipides) developed.

silviculture The cultivation of trees for commercial reasons; forestry.

sima The part of the Earth's crust which makes up most of the ocean floors and also rests beneath the sial of the continents. It consists mainly of silica and magnesium, and is denser than the sial. SEE ALSO **sial**

simulation The use of a model to imitate reality. Humans can use and manipu-

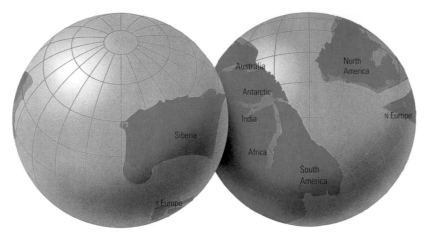

▲ **Silurian** During the Silurian period the distance between the North American and northern European landmasses lessened considerably. By this time all the southern continents had fused together forming the land mass known as Gondwanaland.

late the parts of a model in a fairly subjective manner. Computers can be used in simulation, and they will be objective. Simulations can look into the future and make predictions about what is likely to happen. Students of the **greenhouse effect** often try to construct models of the **atmosphere** to simulate the present-day situation, and then run the models to see what is likely to happen to the world's climate in the future.

sink SEE **sinkhole**

sinkhole A hollow or hole in an area of limestone where solution has enlarged a **joint**. There may have been some collapse of surface material. Water may seep down a sinkhole after periods of heavy rain, but there will not be a permanent flow of water down the hole. Some sinkholes are large enough for spelunkers to descend. ALSO CALLED **sink**. SEE ALSO **swallow-hole**

sirocco A warm, dry wind which blows northward from the Sahara Desert toward the Mediterranean. It generally blows ahead of a **depression** passing along the Medi-

terranean, and can last for one or two days. It is most frequent in spring when the depressions are most active, and the Sahara has already become very hot. The sirocco often has a low relative humidity; it can cause plants to shrivel, and is unpleasant for humans. Sometimes it carries large quantities of dust from the Sahara. If it crosses the Mediterranean, it will pick up moisture and bring hot, humid and rather unpleasant conditions to Sicily and southern Italy. There are occasions when Saharan winds blow as far as northern Europe, where they may cause **thunderstorms** because of the heat they bring. When it rains, the Saharan dust is washed to the ground as **red rain**.

site SEE **position**

situation SEE **position**

sketch map A rough map drawn to show specific information. Generally not very accurate, it does not need to show a wide range of information, such as might be found on an atlas map.

sketch section A cross-section which is designed to show certain specific information. It does not have to be as accurate as a true cross-section, but will be used to show quite clearly certain features of landscape or perhaps land use.

slack A hollow between two sand dunes in a coastal region. The water table may be near the surface of the slack, and it will be inhabited by very different types of plants and animals compared with those living in the very dry conditions on the sand dunes.

slag The waste products which result from smelting iron.

slash and burn The method used by shifting cultivators. They slash down the trees and some of the smaller vegetation, and then set fire to it, in order to clear an area for planting crops. Burning the vegetation creates a layer of potash, which is a useful fertilizer. The fire will kill some pests, but it also destroys useful animals and part of the **humus** layer.

slate Metamorphosed clay or shale. It is normally dark green, black or blue in color, and splits easily because it has well-developed cleavage. It divides up into smooth pieces, which can be used for roofing material. Many old slate workings have closed, because tiles are now used instead of slate.

sleet A combination of rain and snow.

slip-off slope A gentle slope found on the inside of the bend in a **meander**. It is the slope which leads up to a **spur** protruding from the side of the valley into the plain. It is on the opposite side of a river from a **bluff**, and is likely to be an area of deposition rather than erosion. As the river slows on the inside of the bend, deposition of silt or sand may occur, which gradually builds up the slip-off slope.

slum An area of very run-down, poor-quality housing, in or near an urban settlement, usually characterized by overcrowding, multi-occupancy and poverty. There are low levels of sanitation and high levels of garbage.

slum clearance The tearing down of poor-quality housing and the subsequent redevelopment of the area. This may then involve the relocation of the former population elsewhere.

slump A type of mass movement in which rock and soil move downhill. Rocks shear away from a hillside or cliff with a rotational movement, leaving a scar on the rock face. Slump (or slumping) is most likely to occur when rocks are tilted, and when **impervious** clays or shales are overlain by **permeable** rocks.

slurry A mass of wet mud, which is likely to slip down a slope. Slurries can be found in soils such as shales and some clays. They can also occur in man-made conditions, as on dumps and tips of coal waste in mining areas. The Aberfan disaster in Wales in 1966 (when 144 people were killed) was the result of slurry rushing downhill onto the town.

smallholding British name for a small area of land used, generally intensively, for farming purposes. Smallholdings are often located near cities, and the work is done by the owner and his or her family, sometimes on a part-time basis, as a second job. Fruit, vegetables and flowers may be grown, and there may be pigs and poultry. Glasshouse cultivation may also take place. Smallholdings are common in many European countries, such as Belgium, Britain, France and the Netherlands.

smelting The production of metal by melting the ore and extracting the impurities. Some mineral ores such as iron may be as much as 65% pure, and so the waste products are less than the weight of the mineral extracted. In other cases, such as copper or tin, the ores are less than 10% pure, and may even be as low as 1%, and so there are vast quantities of waste products.

smog A mixture of smoke and fog, which used to be very common in London, England. Since the Clean Air Acts of the 1950s, London has been much cleaner, and smogs are rare. Smogs occurred when anticyclonic conditions caused **radiation fog** to develop. If the air remained calm, the fog could get steadily worse, day after day. More soot and industrial grime would also accumulate, as it was unable to escape into the atmosphere, and pea-soup fogs thickened until visibility was often less than 30 ft [10 m]. Other industrial cities had similar problems, especially if they were in hollows, which accentuated the accumulation of fog and nuclei. Pittsburgh in Pennsylvania, USA, is a notable town which has been cleaned. One of the largest problems today is in Los Angeles, California, where the pollution is the result of the large number of automobiles. Los Angeles is in a large basin, noted for inversions of temperature, which create ideal conditions for the formation of fogs and smogs.

snout The downstream end of a glacier. It will be an area of melting for much of the year, and a river is likely to be flowing from underneath the ice, often through a cave. Masses of morainic debris may make the end of the ice look very dirty, and there will be depositional material all around the snout, and also further down the valley, if the glacier is in a period of retreat.

snow A type of precipitation consisting of water vapor which has frozen into ice crystals. Several ice crystals join together to form a snowflake, which will gradually fall down to earth. Some snowflakes may melt as they fall, especially if the temperature is only just down to freezing point. In such conditions **sleet** occurs, and for the same reason it is possible for clouds to give snowfall on mountains, while producing rainfall on adjacent lowland areas. Snow is melted before being measured in a rain gauge, and on average 4 in [100 mm] of snow represent 0.4 in [10 mm] of rainfall. This varies slightly, depending on the type of snow that has fallen. Some snow can be dry and powdery if it has come from a dryish area, but other snowfalls can be wetter if their source is over the sea. Interior regions, such as the **prairies**, generally have dry snow, which is blown about in blizzards; it can be moved off roads by a snow blower. Wetter snow occurs in New England, which requires a snow-plow to move it off the roads.

snowfield An accumulation of snow in a mountainous area. If there is enough

snow to compress the snowfield, it will gradually turn into ice, and eventually **glaciers** will be able to form.

snow line The line on a mountainside above which there is perpetual snow. There may be snow below the line during the winter months, but it will not persist throughout the year. The altitude of the snow line varies with latitude. At the Equator it is about 15,000 ft [5,000 m]; for example, in the Andes of South America or on Mt. Kilimanjaro in Tanzania. In the Alps, the snowline is about 10,000 ft [3,000 m], but it varies from north to south, and there is further variation depending on the aspect of the slopes. In northern Norway, round the Arctic, and in the Antarctic, the snow line is at sea level; for this reason, glaciers flow into the sea to form **icebergs**.

socialism A social and political system based on the concept of common ownership of the means of production. It usually involves a high level of state ownership, and can be seen as the first step toward **communism**; but it does allow some private ownership and democratic values.

softwood Timber which is obtained from coniferous trees. The main use of softwood is in the manufacture of paper, by turning the wood into pulp. Some softwood is used as sawn timber for building houses and furniture, and fragments are used in making chipboard. Most trees planted by forestry organizations are coniferous. This is partly because of the demand for softwood, but also because softwood trees reach maturity far more quickly than hardwoods.

soil The upper layer of loose material resting on top of the rock which makes up the surface of the Earth. It consists of tiny particles derived from the broken-down fragments of rock, together with accumulations of plant remains. The organic remains provide the **humus** and the inorganic particles provide vital minerals. The depth of soil varies from a few inches in some arid regions and on mountainsides to several feet in areas of temperate grassland, temperate deciduous forest and some tropical areas. The soil contains pores, in which air and water can be retained. There are many types of soil, and various methods of classification. A common classification is into **zonal soils** (for example, **podzols**), **azonal soils** (for example, **alluvium**) and **intrazonal soils** (for example, **peat**). SEE ALSO **pedocal, pedalfer**

soil conservation The protection and preservation of the soil, so that it will not be ruined or eroded. Soil is slow to form, but if not managed sensibly it can be lost in a very short time. In order to conserve the soil, it should not be left bare for too long, otherwise wind and rain may start to erode it. Plowing, overgrazing or removing trees can expose soil to erosional forces. If soil is used repeatedly for the same crop, certain minerals can be exhausted, and the soil may be ruined. Once it has lost some of its component parts it breaks up more easily and can be washed or blown away. Soil exhaustion is a major contributor to **soil erosion**. Soil can be conserved by planting grass or trees to protect it, or by planting windbreaks of trees around an exposed area. Runoff should be slowed down, and trees and their roots are vital for this task. In many areas, hillsides are terraced in order to prevent the soil being washed away down the hillside.

soil creep The slow downhill movement of soil particles. The movement can take place if the gradient is as slight as 2°. It is generally so slow that it cannot be seen, but over long periods large quantities of soil and rock fragments slide to the bottom of hillsides. The movement is helped by the effects of lubrication, and soil creep is much more active in wet areas. SEE ALSO **creep, mass movement, solifluxion**

soil erosion The removal of soil from an area by means of wind or water. Soil erosion is most likely to occur where the soil has been left bare. This does not happen very often in natural areas, but it is found in many locations which have been farmed. Erosion may be sheet or gully, and may be caused by wind or water. Wind and water erosion are likely to occur at the same time in the same places. Areas which have suffered from severe erosion include the **dust bowl** on the High Plains of the USA, many areas in the **Sahel** in Africa, the mountains of southern Italy and the uplands of New Zealand. SEE ALSO **gully, sheet erosion**

soil profile SEE **profile** [def. l]

soil texture The size or arrangement of the particles which comprise a soil. They may be fine and small in the case of clays, but are larger and coarse in the case of sands. Often they will be a mixture, called a **loam**.

solar power Heat, energy or electricity derived from solar radiation. It is a source

SOLAR POWER

In active solar space heating, an example of solar power, a large collector (1) is needed and air (2) is heated directly by sunlight (3). The hot air then returns to a rock storage tank, where its energy can be stored until needed (4). Valves (5) control the flow of the heat to the living areas of the house (6). The flow is controlled by a pump (7).

of energy which has not yet been exploited to the full, although undoubtedly it will be one day. It is already used in many places on a small scale. There are many houses which have solar panels to create heat for water supplies or central heating. There are solar-powered irrigation pumps in use in some parts of the **Third World**. Solar power is most likely to be useful in tropical areas, which have long hours of sunlight. Deserts often average up to 3,000 hours of sunlight per year.

solfatara A vent or small crater in the surface of the Earth which emits steam and a variety of gases. Many of the gases are sulfureous. Solfataras are thought to be signs of dying volcanic activity, and they occur mostly in volcanic areas. Solfatara (derived from the Italian *solfa*, meaning "sulfur"), an area near Naples, gave its name to the phenomenon, and there are other examples in Italy, as well as in Hawaii, New Zealand and elsewhere.

solifluction SEE **solifluxion**

solifluxion Soil flow which is slightly faster than **soil creep**, though still very slow and not visible to the naked eye. It is the movement of soil and rock particles down slopes as a result of saturation, followed by freezing and thawing. As the water freezes at night and then thaws during the day, particles of soil are moved and then slip downhill because of the effects of gravity. Both water and ice make these movements easier by means of lubrication. Solifluxion is most likely to occur in **tundra** areas. ALSO CALLED **solifluction**. SEE ALSO **mass movement**

solstice Either of two dates in the year when the Sun appears to stand still. Because of the tilt of the Earth's axis and the revolution of the Earth around the Sun, it appears that the Sun is moving. As it moves northward from the Equator in March toward the **Tropic of Cancer**, the northern hemisphere experiences its summer. When the Sun reaches the Tropic of Cancer on June 21, it appears to stop moving,

before returning to the south; this is the **summer solstice**. Similarly, when it reaches the **Tropic of Capricorn** on December 22 it apparently stops again, and this is the **winter solstice**. On the summer solstice the Sun is directly overhead at the Tropic of Cancer, and it is the longest day of the year for the northern hemisphere. At the winter solstice, it will be the shortest day of the year in the northern hemisphere.

solution A form of chemical **weathering**. It is particularly active in limestone areas where the **joints** can be enlarged to form **grikes** and **potholes**. Solution can also be active in chalk areas and anywhere with rocks containing salts.

South, the The name used by the Brandt Commission to describe the poorer countries of the world, as opposed to **the North**, which includes North America, Europe and Australasia. Most of the South is in tropical and subtropical latitudes, and includes large areas of South America, Africa and Asia.

sovkhozy (*Russian*) A state farm in the former Soviet Union, set up prior to the political changes of 1991. It differs from a collective in that it would have been set up by the government in an area which had not been fully farmed in the past, or in an area with particular problems. For example, many parts of European Russia near to the Baltic have large areas of glacial **drift**, which are not very fertile. Drainage and reclamation schemes were set up by the government, special equipment was provided, and financial assistance was given to help with the management of the land. State farms in the Arctic conducted research into growing crops in the very short summers. When the dry steppes were plowed in the 1950s, there were no farms there, and so state farms were created. They had both manager and labor supply brought in from other parts of the Soviet Union.

space An area of unspecified size. Spatial distributions are very important in geography. Most studies have to consider the way in which space is utilized. The distances between two locations is also important, as this will influence spatial relationships.

spatial analysis An analysis and explanation of the distribution of people or an activity. In spatial analysis, attempts will be made to establish which factors have been instrumental in explaining and accounting for the distribution.

spatial distribution The manner in which settlements or other spatial phenomena are set out and the distances between them. It can be explained in a descriptive way, but it will also need to be quantified if possible, by means of mapping, or by use of the **nearest neighbor analysis**.

Spearman rank correlation The method used by Spearman to compare a paired set of ranked values in order to look for a correlation. The higher the correlation coefficient, the greater is the correlation between the two sets of numbers. SEE ALSO **correlation**

sphere of influence The area around a store, settlement or other central place, which is affected by the services or goods on offer. In the case of a city, the sphere of influence includes all the surrounding countryside and towns from which customers come to the stores, schools, hospital, etc. A small town has a small sphere of influence, whereas the sphere of influence of a large city such as London or Paris may extend for hundreds of miles. SEE ALSO **central place theory**

spit A depositional area along a coast, which has been built up as a result of **longshore drift**. A spit is attached to the land at one end, and may consist of sand, mud, shingle or any combination of these. Most spits protrude across an estuary, and may eventually block the estuary to form a delta. SEE ALSO **bay bar**

spontaneous settlement A settlement which grows up without planning,

and can often appear very quickly, such as a **shanty town**. At first, the housing is of very poor quality, although this would gradually improve if the settlement became permanent. Most spontaneous settlements are the result of homeless people making use of an empty piece of land, often in a rather undesirable location. ALSO CALLED **squatter settlement**

spot height A height marked on a map at a particular point. Spot heights are often located on hill tops.

spread effect The filtering through of wealth from a richer, core area to poorer, peripheral ones. For example, an increase in economic activity in a central area might increase the demand for raw materials from peripheral regions.

spring A point on the Earth's surface at which underground water emerges. It is generally located where there is a change of rock type, and the water flows out at the top of an **impervious** layer, such as slate, shale or clay. There are many springs at the foot of limestone and chalk hills. Spring water is often very clean and pure, though it will normally contain large quantities of calcium in solution.

spring equinox SEE **equinox**

spring line A series of springs along the foot of a chalk or limestone hill, which may give rise to farms or even villages. The spring line generally follows the junction of the limestone or chalk and an **impervious** rock which makes up the lowland area adjacent to the hills.

spring tide SEE **neap tide**

spur An area of high ground which protrudes into an area of lower ground, such as a river valley. Spurs alongside valleys are gradually eroded by the river, but, in the upper and middle courses especially, they often form **interlocking spurs**. In glacial valleys, many of the spurs will have been truncated.

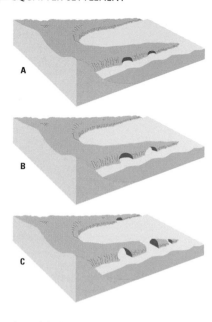

▲ **stack** As the peninsula is eroded by the sea, at first caves start to appear the sides of the peninsula (A). The peninsula becomes narrower and the caves become larger until arches form (B). As the arches grow further they collapse, leaving stacks.

squatter settlement SEE **spontaneous settlement**

stability of air A state of equilibrium of the air in the atmosphere, which occurs when its lapse rate is less than the **dry adiabatic lapse rate**. This means that if the air is moved upward, it will be cooler and therefore heavier than the surrounding air, and it will gradually sink down again to its original level. Subsiding or stable air will not give any precipitation. Stable equilibrium can also be achieved when inversion of temperature has taken place. SEE ALSO **temperature inversion**

stable population A population where fertility and mortality rates are constant; such a population grows at a constant rate and has an unchanging age distribution. If the annual growth rate is 0, the population is seen as **stationary**.

stack A small rocky islet, generally located at the end of a peninsula. It is usually a result of **erosion**, which opened up joints in the peninsula to enable **caves** to form. As the caves grew larger, arches would develop, so that eventually a break could open up in the peninsula. When this happened, the end of the peninsula may have been left as an isolated remnant – a stack. There is often more than one stack at the end of a peninsula, evidence of considerable erosion over thousands of years.

stalactite An icicle-shaped deposit, found hanging from the roof in an underground cave. Usually made of calcium carbonate, stalactites are formed by water slowly percolating through the rocks; as the water droplets are about to fall from the cave roof, a tiny layer of calcium precipitates out of the water and solidifies. Gradually, the deposits build up to form a column of calcium carbonate suspended from the roof. SEE ALSO **stalagmite**

stalagmite A column of calcium carbonate which grows upward from the floor of a cave in a **Carboniferous** limestone area. It is formed as a result of water dripping from the ceiling of the cave and leaving tiny deposits of calcium on the floor. The deposits gradually build up, and can sometimes extend upward to meet the stalactite above. Stalagmites tend to be fatter than stalactites, possibly because the water splashes out when it drops to the floor, causing a wider spread of calcium. Stalagmites and stalactites probably grow at a rate of 0.4 in [10 mm] per 100 years on average.

standard of living 1 The conditions for living thought to be necessary, as laid down by national or international agreements; for example, minimum wages, maximum working hours. **2** The conditions of living which people do have or would like to have.

standard time A system of time used for a wide area, instead of local time for lots of smaller areas. This is usually **Greenwich Mean Time** or a time differing from it by a particular number of hours.

staple diet A main and basic foodstuff. In North America and Europe, bread is a basic food eaten by almost everyone, but most people have a fairly varied diet. Many people in Southeast Asia will have rice as their staple foodstuff, and in many parts of Africa corn is the main food. Tropical forest inhabitants may eat cassava as their basic food, and some herders may be dependent on milk from their animals and live on a diet of cheese or yogurt.

steppe An area of temperate grassland found in Kazakhstan, Russia, Ukraine, and in neighboring parts of southern Europe and in the Danube valley. The landscape is often very flat and open, and the climate is too dry to support many trees. When plowed up, the grassland can become very rich farm land, especially for growing **cereals**. Steppes are very similar to **prairies**, and have a similar continental type of climate. Winters are very cold, well below freezing point; but the summers are quite warm, 60°F to 70°F [15°C to 20°C]. Rainfall is quite light, about 20 in [500 mm], and falls mainly in summer. The winters are dry, with only light falls of snow.

Stevenson screen A white box on legs, designed to house weather instruments. It

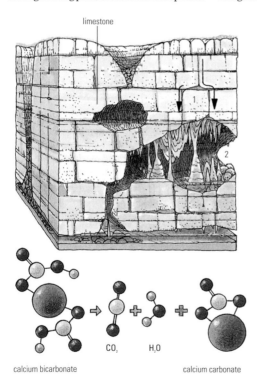

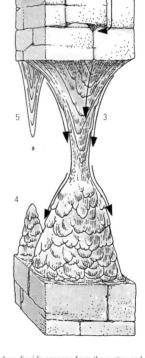

▲ **stalactite/stalagmite** Weak carbonic acid, which is formed when rain absorbs carbon dioxide from the atmosphere, dissolves tiny amounts of calcium carbonate from limestone rock as it percolates down through the rock (1). If the water enters a previously eroded cave (2), it forms drips (3). Some of the dissolved carbon dioxide escapes from the water, and the calcium bicarbonate in the water is precipitated as a calcite known as travertine, consisting of calcium carbonate. The travertine forms rising columns called stalagmites (4) and descending pillars from which water drips, called stalactites (5).

stands about 4 ft [1.3 m] above ground level, and has louvered sides to allow free circulation of the air. It is painted white in order to reflect heat. It will contain wet and dry bulb thermometers, and probably maximum and minimum thermometers too. It may also have a **thermograph** and a **barograph**. All weather stations have a Stevenson screen, as it is the standard method for obtaining temperatures which are unaffected by sunshine and strong winds. SEE ALSO **hygrometer**

stock A granite intrusion which is like a small version of a **batholith**.

Stone Age The period when early humans were using and shaping objects from stone. They often made use of **flint**. The Stone Age can be divided into the Paleolithic (or Old Stone Age), the Mesolithic (or Middle Stone Age), and the **Neolithic** (or New Stone Age), as the people gradually improved their implements and techniques. The Stone Age ended in Europe around 2000 BC; it ended later in North America, and only quite recently in parts of Papua New Guinea.

storm A period of strong winds, which is particularly noticeable at sea.

storm surge A rise in the sea to a higher level than expected as a result of strong winds. It is most likely to be associated with the time of high tides, and is likely to cause flooding. It was a storm surge in the North Sea which caused the severe flooding in eastern England and the Netherlands on January 31, 1953.

stratigraphical column SEE **geological column**

stratigraphy SEE **stratum**

stratocumulus A layer of cloud which also contains some areas with vertical extent. The vertical patches of **cumulus** may cause heavy showers of rain. If numerous cumulus clouds merge into one large sheet, they are known as "cumulostratus."

stratopause The boundary between the **stratosphere** and the **mesosphere**, at about 30 miles [50 km] above the Earth's surface. SEE ALSO **atmosphere**

stratosphere The zone of **atmosphere** which extends from the **tropopause** (at about 10 miles [15 km] above the Earth's surface) to the **stratopause**.

stratum A layer of **sedimentary rock**, usually one of a series of layers which have been deposited one on top of another, as, for example, on the seabed. Each stratum will indicate the conditions which prevailed at the time of its deposition. The study of the strata is called "stratigraphy," and the stratigraphical or **geological column** contains all the periods of geological time. Rock strata may be very thin, of just a few inches, or they can be several feet in extent. Initially, they are usually horizontal, but they are likely to be affected by folding and faulting.

stratus A low horizontal sheet of **cloud**. A thin layer of stratus is light gray in color and will give no rain. Thicker stratus clouds are darker, and may give light or even steady rain. Stratus clouds often cover mountains, while neighboring lowlands remain cloud-free. Stratus clouds are often found in warm sectors of **depressions**.

stream order A classification of streams in a drainage basin through numbers, mostly by the method developed by A.N. Strahler in 1957. First-order streams are all outermost streams with no branches; when two first-order streams meet, a second-order stream is formed, and when two second-order streams meet, a third-order stream is formed. This continues down the channels until the main river, or opening to the mouth, is reached.

striation The process in which scratches and grooves ("striae") are left by ice scraping over rocks. It is the small rock fragments in the ice which probably do most of the damage. The marks show the direction of the ice movement, and can be seen in many places where there has been recent ice activity.

strike The direction of a horizontal line along a rock stratum, which is at right angles to the **dip** of the rocks. The relationship between the land surface and the **contours** is similar to that between the dip and the strike.

strip cultivation A method of farming which helps to conserve the soil. On sloping ground, different crops are planted in strips to ensure that the land is not all harvested and bare at the same time. The strips often follow the lines of the **contours**, and grass and a cereal are frequently alternated down the slope. Strip cultivation is an effective way of preventing the soil being washed away.

strip development Urban expansion along the roads leading from large towns. A negative aspect is the replacement of fields and other green areas with parking lots.

strip grazing The subdividing of a field into small strips so that only one part is grazed at a time. The animals, generally cattle, eat the grass in one strip, and are then moved on to the next, which will be bounded by an electrified wire fence. As they progress, they leave behind rich deposits of manure. Strip grazing is practiced in New Zealand, parts of Britain and in the Alps.

subaerial erosion Erosion on the surface of the Earth.

subcrustal convection currents Convection currents in the **mantle**, which are responsible for the movement of the plates on the Earth's crust. The currents are caused by the heat present within the Earth.

subduction zone The downward movement of rocks and crustal material at a zone of convergence. When two plates are moving together, it is inevitable that one goes beneath the other, and this is the zone of subduction. It is normally indicated by a trench, an island arc, or a region of volcanic activity, such as near the Philippines and Japan. SEE ALSO **plate tectonics**

subglacial Beneath a glacier or ice sheet.

submerged coast A coastline formed when the sea level rose and submerged the former area of land. Such features as **rias** and **fjords** may be found in these areas.

submersible A small craft for underwater exploration, research or engineering. Some submersibles carry a human crew. Others carry only instruments and are operated by remote control from the surface. Submersibles without a crew are generally smaller and can can resist water pressure at greater depths.

subsequent stream A stream which is a tributary to the **consequent stream**. The consequent stream would have been the main stream to develop on a slope. Subsequent streams often flow at right angles to the consequent, and generally follow the **strike** of the rocks. They often flow along the softer and weaker rocks and may carve larger valleys than the consequent streams.

subsidence A sinking to a lower level, either of part of the Earth's crust or of air. If the crust subsides, it may be on a large scale (for example, causing a **rift valley**) or on a small scale (for example, the collapse of land in mining areas). Air subsiding involves a large air mass sinking slowly downward, usually warming as it goes.

subsistence crop A crop grown by a farmer entirely for his or her own consumption. SEE ALSO **cash crop**

subsistence farming Farming in which the produce is consumed by the farmer and his or her family. The system produces little or no surplus for sale, unlike commercial farming. Subsistence farming is still found in isolated areas in advanced countries, but it is more characteristic of **Third World** countries.

subsoil The layer which is situated below the **topsoil**. It may not contain any organic matter, but is likely to include rock fragments. It is roughly equivalent to the **C horizon**.

suburb A built-up area on the outskirts of a city, generally consisting of housing developments built in the 1930s or after World War II. Some suburbs may have their own shopping areas, which are mini **central business districts**.

summer solstice In the northern hemisphere, June 21; but December 22 in the southern hemisphere. SEE ALSO **solstice**

sunrise/sunset The first and last appearances of the Sun above the horizon,

▼ **submersible** Modern submersibles are versatile vehicles used for scientific and industrial purposes. Most can be fitted with a wide range of equipment, such as video, still cameras and claw-like manipulators. Some types, known as diver lockout submersibles, such as the Johnson Sea Link shown here, are equipped with pressure chambers from which divers can enter and exit, allowing them to perform work and then return to the submersible for gradual decompression.

where the Sun seems to rise or set. The times of each of these vary throughout the world with latitude and depending on the declination of the Earth due to its rotation on its axis.

sunshine recorder An instrument for measuring the amount of sunshine. The standard sunshine recorder is the Campbell-Stokes recorder. It consists of a glass sphere, which concentrates the rays of the Sun and focuses them on to a piece of specially graduated paper. The rays makes scorch marks on the paper, and this can be measured to give the length of time that the Sun was shining.

sunspot A dark area on the Sun's surface, which gives out a higher level of solar radiation. The number of sunspots varies, but seems to reach a maximum every 11 years or so.

supercooled A term used to denote drops of water which have been cooled below freezing point but have remained in liquid form. Supercooling can occur if the air is very still. The supercooled droplets freeze on to objects moving through them, such as the leading edge of aircraft wings,

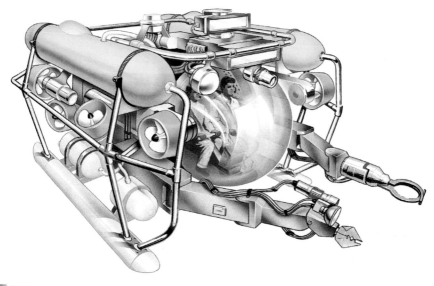

and if they drift with a very gentle air movement they may freeze on to the sides of buildings, trees or telegraph poles.

superimposed A term used to denote a river system which does not follow the trend of the rocks over which it is flowing; this is because it is continuing to follow the trend of other rocks which have since been eroded.

surface runoff The portion of rainfall which runs off either as channel flow or as overland flow, but does not percolate into the soil and rocks, and is not evaporated into the atmosphere. ALSO CALLED **immediate runoff**. SEE ALSO **delayed runoff**

surge A large or massive movement of water, generally in the sea. SEE ALSO **storm surge**

suspension The state in which small particles of sediment are carried by rivers. The particles are held up in the water by turbulent upward eddies, which prevent the tiny particles from falling to the bed of the river. It is the suspended particles which make rivers look muddy. It has been estimated that, in a year, about 200 tons of solid matter are removed from each square mile of the Earth's surface; so there is a lot of material for the rivers to carry. In the USA, the River Mississippi carries some 300 to 400 million tonnes in suspension each year.

swallow hole A large **pothole** in a limestone area, down which a river is flowing. The hole will be an enlarged **joint**, which has been weathered by **solution** and also eroded by running water. Some swallow holes descend for up to around 300 ft [100 m]. SEE ALSO **sinkhole**

swash SEE **backwash**

symbiotic relationship The mutually beneficial relationship between two different organisms. This term has been extended to describe even such relationships as between a city and its surrounding area, and between humans.

syncline The downfold in folded strata. The upfold is called the **anticline**. The rocks in the syncline are often compressed, which makes them harder and more resistant to erosion, and they often erode more slowly than the rocks in the anticline. SEE ALSO **inverted relief**

synoptic chart A synopsis or summary of the weather which is drawn on a map each day. The chart shows the exact meteorological conditions at a certain time, with isobars, weather systems, temperatures, wind speeds and directions, amounts of cloud, etc.

synthetic fiber A fiber which does not occur naturally and is manufactured by chemical synthesis from wood, coal or petroleum. The earliest synthetic was nylon, made from coal and developed commercially in the 1940s. In the 1950s, rayon was made from wood, and since then there have been many discoveries and developments, mostly using petroleum as the raw material. Synthetics have special qualities; they may be drip-dry, crease-resistant and hard-wearing, but none of the synthetics has quite replaced all the qualities of wool and cotton. ALSO CALLED **artificial fiber**, **man-made fiber**

system Any set of related elements. This can consist of only abstract parts, but is usually used as a concept by geographers to look at features in different areas, such as **ecology** or **geomorphology**. A system consists of inputs, outputs and throughputs, and may be studied at any scale, thereby sometimes making it difficult to mark out boundaries. The use of systems is very useful in geography to study how different parts of the landscape, natural and man-made, function. SEE ALSO **open system**, **closed system**, **systematic geography**

systematic error An error which occurs repeatedly in samples, usually in a particular direction, and will therefore cause the results of the survey to be biased.

systematic geography That part of geography which deals with the overall study of a particular subject, such as agriculture, transportation or industry, rather than the study of these subjects within a particular region. It is an approach which can give a fuller understanding of the topic or system under consideration. For example, a glacial system is a glacier, together with all its inputs and outputs. If one specific glacier has been studied, it is likely that other glaciers will be understood to a significant degree because of the similarities. Any geographical subject can be considered as a **system**; for example, an urban system – the study of one city in detail, will help to explain how all towns have developed and function.

T

Taafe A model named after E.J. Taafe, who made a study of the growth of communications in an African country. There are various stages of transportation development. Each tiny port has a small hinterland served by paths and trackways. A road or railroad may be built to link an inland mine or estate with the port. Inland centers which grow quickly will be linked with other inland towns, and the major port or ports will gradually build better roads or railroads to the inland towns.

taiga The coniferous forest which extends over thousands of square miles in Siberia, Russia, Finland, Sweden, Norway and Canada. The climatic conditions are unsuitable for deciduous woodland, because the growing season is seven months or less, but coniferous trees can survive. Summer temperatures are around 60°F [15°C], and January averages 15°F to –5°F [–10° to –20°C]. There are several months with temperatures below freezing point, and the **subsoil** is frozen for much of the year. The roots of the coniferous trees spread out horizontally as they cannot penetrate vertically into the ground. The main types of tree are spruce, larch, fir and pine. In the southern parts of the taiga, the trees grow to 50 ft [15 m] in height, but further north the trees become smaller until they are little more than bushes, and tundra vegetation can be seen. The trees are a source of **softwood** for pulp, which is used to make paper. Thousands of trees are cut down every day, but in most countries reforestation ensures that the trees are replaced. Many highlands such as the Alps have coniferous forests which are similar to the taiga. Wherever coniferous forests occur, the needles fall to the ground to produce **podzols**, which are acid and ashy gray in color. SEE ALSO **coniferous forest**, **deciduous woodland**, **evergreen trees**

take-off point Stage 3 of **Rostow's model**. All advanced countries have gone through the various stages of the Rostow model, and take-off began at different times in different countries. In Britain, take-off occurred about 1800, in France and the USA it was around 1860, and in Japan about 1900. In Brazil, it occurred in the 1950s. SEE **Rostow's model**

talus An accumulation of small rock fragments which formed on the side of a hill and then slid down to its base, because of the effects of gravity and lubrication by water. The rock fragments are the result of **weathering**, especially **freeze-thaw** activity, and are generally referred to as **scree**. Talus commonly consists of angular fragments of rock, which if they were cemented and consolidated would be called **breccia**. When they slide downhill, they may accumulate on the valley floor in a triangular or cone shape, rather like a rock delta. All types of rock can form talus.

talweg (*German*) SEE **long profile**

tank A lake formed by building a dam across a stream or an embankment around a field which has been covered by flood water. Water collects in the tank during the rainy season, and can be used for growing crops when the rainy season has finished. Such supplies of water will not last throughout the year, but they are retained for a few extra weeks. As the water level subsides, crops are planted in the saturated soil. Tanks are found in many parts of India.

tanker A ship, aircraft, road or rail vehicle designed to carry fluids, especially oil, in bulk. Road tankers carry oil from refineries to consumers, and take milk from farms to dairies. In the 1950s, seagoing tankers were normally vessels of 20,000 or 30,000 tons, but throughout the 1960s the size of tankers increased until 300,000-ton tankers were in use. The large tankers were able to transport oil more economically than the smaller vessels. Because they need very deep water in the harbor, special oil ports grew up to handle the tankers.

tarn A small lake, especially in the Lake District in Britain. Tarns are often the lakes which have formed in **cirques**; for example, Red Tarn on Helvellyn. The name is derived from the old Norse word for a "teardrop."

tear fault A fault in which the movement is along the horizontal. Tear faults occur frequently in areas of tectonic activity. The amount of movement may only be a few feet, but can be several miles in some cases. The San Andreas Fault in California is a well-known example of a tear fault. ALSO CALLED **lateral fault, wrench fault**.

tectonics The movements which affect the features of the Earth's crust. The processes which bend or crack the crustal rocks to cause folding or faulting are the result of tectonics, resulting from movements of the earth's plates. SEE ALSO **plate tectonics**

temperate Having a moderate climate, such as midlatitude areas. Strictly speaking, the temperate zone is the area between the tropical or torrid zone, and the frigid zone; it extends from the **Tropic of Cancer** to the Arctic Circle and from the **Tropic of Capricorn** to the Antarctic Circle. The summers in temperate latitudes tend to be warm, but the winters are cool. Temperate climates are those which have neither very high nor very low temperatures.

temperate grassland SEE **grassland**

temperature Temperature is measured by a thermometer and generally recorded in degrees Celsius or Fahrenheit; 32°F [0°C] is the freezing point of water and 212°F [100°C] is the boiling point of water. The sun is the major source of heat. Temperatures generally decrease from the Equator toward the poles because of the corresponding decrease in the angle of the sun away from the Equator. Temperatures also decrease with height above sea level because the air becomes thinner. Latitude and altitude have both been mentioned as major factors influencing temperature, and the third factor is the difference between land and sea. The land heats up and cools down more quickly than the sea, and so it is generally hotter in summer and colder in winter.

temperature inversion An anomalous increase in temperature with height. Normally the temperature of the air decreases from ground level upward. The average rate of decrease is 1°F for every 90 ft [1°C for every 160 m]. In certain meteorological conditions, the normal situation is reversed. On a clear, calm anticyclonic night, the cool air may roll downhill and accumulate in valleys, and the air temperature will be lower near the valley bottom than it is 300 ft to 600 ft [100 m to 200 m] higher. Above the cold layer there will be warmer air, which is likely to form cloud or haze. On summer evenings, it is often possible to see evidence of a warmer layer if there is smoke rising from a bonfire. The smoke will rise vertically and then bend horizontally when it reaches the "inversion layer." If this situation develops on a larger scale, the dust and dirt rising into the atmosphere are trapped and unable to escape, giving rise to serious pollution, as, for example, in Los Angeles, California. As the weather conditions are often anticyclonic and Los Angeles is situated in a hollow, the fumes from cars cannot escape very easily until the air becomes more turbulent, causing the layers to mix and the inversion layer to disperse.

tephra The ash, dust and cinders thrown out by a volcanic eruption. The term does not include **lava**. When tephra is thrown up into the air by an explosion, it will soon fall again, quite close to the crater. The smaller and lighter particles are blown further away and accumulate on the downwind side of the eruption. Large deposits of tephra were left by the eruption of Heimaey in Iceland in 1973. So much tephra fell on to some houses that the roofs collapsed under the weight.

terminal moraine A moraine formed at the end of a glacier when the ice has reached its maximum extent and begins to melt. Terminal moraines are generally slightly curved in shape. Because of the changing location of the **snout**, several "recessional moraines" may form at the end of a glacier, and if the ice is thick, they can form small ranges of hills. Long Island, New York, in the USA, for example, has a terminal moraine.
SEE ALSO **glacial deposition, moraine**

terracette SEE **sheep track**

terracing A series of terraces cut into a slope in order to create strips of flat land. Land is dug out of the slopes and used to build up new flat areas. Walls or mud banks are used to preserve the level patches and to help retain water if **irrigation** water is being supplied to the terraces. Terracing is used for farming in a variety of regions; for example, in rice-growing areas in Southeast Asia and in wine-growing areas of France. It was also used by the Incas for producing corn on the Peruvian Andes.

terra rossa (*Italian*) Red earth which has gained its color because of the iron con-

tent. *Terra rossa* is a residual soil found in limestone areas of the former Yugoslavia.

terra roxa (*Portuguese*) A reddish iron-rich soil which formed on the basaltic rocks of the São Paulo state on the Brazilian plateau. It is particularly suitable for growing coffee.

territorial seas/waters The coastal waters, together with the seabed beneath and the air above, over which a state has sovereignty (i.e. control over). From the 18th century, this area was seen to be 3 nautical miles [3.5 miles, 5.5 km] from the shore; such areas are often difficult to measure, since many countries have very indented coastlines and therefore a baseline is used. Most countries claim at least 12 nautical miles [14 miles, 22 km]; the 1983 Law of the Sea Convention suggested an Exclusion Economic Zone of 200 nautical miles [230 miles, 370 km], which also included the resources of the water and the seabed. In areas where there is not at least 400 nautical miles [260 miles, 740 km] between coastlines, a median line is drawn between the two baselines.

TERRACING

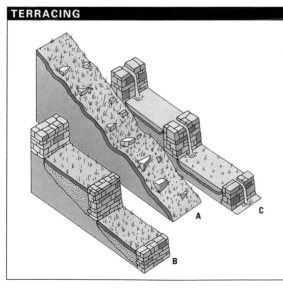

Since prehistoric times, humans in hilly or mountainous areas have used terracing to increase the area of land that is available for cultivation and to help prevent soil erosion (A). Banks or walls are used to retain earth. Often field stones are used for the construction (B). Terraces increase the area of flat or near level land that is available for growing crops, including trees, or for grazing. In Asia, terracing is often combined with complex systems of irrigation in order to raise rice in paddy fields enclosed by the terrace walls (C).

Tertiary The geological period which followed the Primary and Secondary, and preceded the **Quaternary**. The Tertiary was the era of mammals, when modern animals and also modern plants became dominant throughout the world. It began about 65 million years ago, and includes the following geological epochs: Paleocene, Eocene, Oligocene, Miocene and Pliocene. The Alpine **earth movements** occurred during the Tertiary, which formed all the young fold mountains, such as the Alps, Himalayas, Andes and Rockies. There was volcanic activity in southwest Scotland and northeast Ireland, caused by fissure eruptions where North America was drifting away from Europe. SEE ALSO **geological column**, **geological time scale**

tertiary sector The part of the economy which includes all the service indus-tries; that is, the jobs connected with administration, transportation, education, medicine, banking, etc. In highly developed countries, such as those of North America and Western Europe, tertiary industries often employ over 50% of the work force, especially in the large cities.

textile Wool, cotton, flax and silk are sources of natural textiles, and there are many synthetic fibers such as nylon, rayon, acrilan and terylene. The manufacture of cloth has seen a recent decline of natural fibers as synthetics have become cheaper and more versatile. The world's major industrialized nations, such as the USA, Japan, Korea and Taiwan, are the leading suppliers of textiles.

thalweg SEE **long profile**

theodolite Surveying instrument dating back to the 16th century, used to measure horizontal and vertical angles. Its modern form consists of a telescope (with cross hairs in the lens for accurate alignment) mounted to swivel in both directions and levelled with a spirit level.

THEODOLITE

telescope
tripod
laser
spirit levels
upper horizontal plate
lower horizontal plate
horizontal adjusters

A theodolite is a surveying instrument used to measure angles between different points accurately, enabling them to be plotted in three dimensions. The telescope is used to locate measuring points on surveying poles; to align the telescope, it is rotated about a horizontal or vertical axis. The precise angles of rotation in these planes are measured by using a micrometric microscope to read a scale on glass protractors. Modern computerized theodolites take these measurements electronically, and use lasers to measure distance.

theory An organized set of ideas used in explaining events or phenomena, usually supported by fact, but perhaps also including some speculation. Scientific theories are more structured, with a number of steps of observations and assumptions, linked to and supported by each other.

thermal electricity Electricity produced from coal, oil or natural gas. The raw materials are burned to produce steam, which then drives the turbines. Coal-mining areas, such as West Virginia in the USA, have several thermal power stations; the same is true of oil-refining locations, such as Houston in Texas, USA.

thermograph An instrument which keeps a written record of temperature. It consists of a thermometer made up of two strips of metal, which expand and contract with changes of temperature, and this is attached to an arm with a pen, which records on to a special piece of paper attached to a revolving drum. A clockwork mechanism inside this drum turns it around once in a week. At the end of the week the paper can be taken off, and it will show the temperature changes of the previous week. A thermograph is not as precise as a thermometer, but it shows all the changes and trends quite clearly.

thermometer An instrument for measuring temperature. There are various types, such as maximum, minimum, grass minimum, and wet and dry bulb thermometers. Some contain mercury, while others contain alcohol; they both work efficiently. Temperatures can be read off a scale on the glass tube.

thermosphere The zone in the **atmosphere** above the **mesosphere**, with a range of about 50 miles to 60 miles [80 km to 100 km] in altitude.

Third World Those countries, especially in Africa, Asia and Latin America, which are aligned with neither the developed nations of the West nor the members of the communist bloc. The expression is rather unsatisfactory as it includes countries which are quite different from one another in many respects. SEE ALSO **First World, Second World**

thorn forest A type of thorny scrub found in areas with prolonged periods of dry weather. In order to survive the periods of drought, the plants often have thick bark, or long roots, and are covered with thorns.

threshold l SEE **sill** [def. 1]. **2** The end of a **fjord** at the point where it enters the open ocean.

threshold population The minimum number of people in a region required to support a particular store or service.

throughflow The flow of water down a slope through the soil. This often happens when the quantity or rate of water falling

THERMOMETER

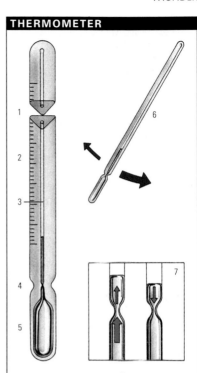

A clinical thermometer is an ordinary mercury-glass thermometer with a particularly fine capillary (3). A constriction (4) allows mercury to flow easily from the bulb (5) but, by surface tension, prevents flow back (7). A temperature reading can thus be maintained on the scale (2) until the mercury is forced back by shaking (6). For easy reading the stem is lens-shaped, as in cross-section (1), to visually magnify the mercury when the thermometer is held at the correct angle.

on the surface of the ground is too much to be all absorbed quickly downward.

throw SEE **fault**

thunder The loud noise which accompanies a flash of **lightning** and is caused by violent disturbance of the air by an electrical discharge. Because the speed of light is faster than the speed of sound, the lightning appears to precede the thunder, unless it is directly overhead. The delay between lightning and thunder is a rough indication of how far away the storm is.

thunderstorm A storm caused by convectional activity and accompanied by thunder and lightning. Because of the size of the convection currents, some thunderstorms give very heavy rain and can cause serious flooding. Equatorial areas receive convection rain and may have a thunderstorm nearly every day. Prolonged thunderstorms develop when there is a large supply of moisture, so that giant **cumulonimbus** clouds can form, and a high **lapse rate**, which will ensure a rapid and persistent upward movement of air for 10,000 ft to 13,000 ft [3,000 m to 4,000 m] above the base of the cloud. If the drops of water are forced so high that they freeze, there will be **hail** as well as heavy rain. SEE ALSO **convection rainfall, lightning, thunder**

tidal limit The highest point in a river inlet which is ever reached by the sea.

tidal range The difference between the water level at high tide and the water level at low tide. In some places in the Mediterranean, the tidal range is only a few inches, whereas the Bay of Fundy in eastern Canada has a tidal range of 50 ft [15 m]. SEE ALSO **neap tide**

tidal wave SEE **earthquake**

tides The periodic rise and fall of the sea caused by the pull exerted on Earth by the Moon, and, to a lesser extent, by the Sun. In many parts of the world there are two high tides and two low tides every day. The

time of each high tide is 12 hours 20–25 minutes later than the preceding tide, because the position of the Moon relative to the Earth will have changed by a small amount after 12 hours have elapsed. SEE ALSO **neap tide**

tidewater (*of a location*) Situated close to tidal water, either on the seashore or in a tidal estuary. Tidal water enables bulk-carrying vessels, the cheapest form of transportation, to penetrate further inland, as, for example, in London, England.

tierra caliente (*Spanish*) The hottest parts of tropical Central and South America, which are the low slopes of the Andes found near the Equator – on the coastal plains and in the lower Andes at altitudes of up to about 3,500 ft [1,000 m]. In the *tierra caliente*, tropical forests can grow because there is a constant temperature of about 80°F [25°C] and up to 100 in [2,500 mm] of rainfall spread right through the year. Cassava is the major food crop, and bananas, sugar cane and cocoa are grown.

tierra fria (*Spanish*) The cold areas of tropical Central and South America, which are found higher up in the Andes, at about 6,500 ft to 10,000 ft [2,000 m to 3,000 m] in equatorial latitudes. The average monthly temperature is about 60°F [15°C] throughout the year. The total

▼ **tides** It is primarily the gravitational pull of the Moon that causes tides. However, pariticularly high (spring) tides occur when the gravitational pulls of the Sun and the Moon are in alignment with each other, and particularly low (neap) tides when they are entirely out of alignment with each other.

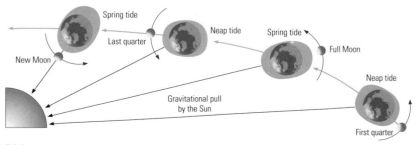

▶ **tides** The power of tides can be harnessed to generate electricity. Tidal power uses a barrage (1) across an estuary or bay. The barrage contains turbines that can spin with a flow of water in either direction. As the tide comes in, gates on the barrage remain closed until a head of water has built up on the sea side of the structure (2). The gates are then opened (3), and the incoming tide flows through the barrage driving the turbines (4). As the tide falls, the process is reversed with the gates closed until the sea has fallen below the level of water retained in the estuary (5). A second, similar method of utilizing the power of the sea harnesses wave power (6). The key difference is that the turbine (7) is air driven, not turned by water. As a wave hits the shore, the force of the water (8) drives air (9) through the turbine blades (10). When the water level drops, air is sucked back down through the turbine spinning it again.

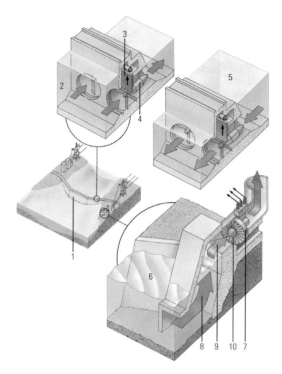

rainfall is spread throughout the year, but is no more than 40 in [1,000 mm]. The rainfall in equatorial latitudes is convectional, which is caused by heat. As the temperature decreases with height, there is less heat and therefore less convection. For this reason rainfall decreases with height in equatorial areas, in contrast to temperate latitudes, where rainfall increases with height. In the *tierra fria*, wheat is grown and barley, too. Potatoes are a staple food and also a source of alcohol. Sheep and llamas are reared. The natural vegetation is coniferous forest in the lower parts, with scrub and grassland higher up. In the *tierra fria*, climatic conditions are preferable to those in the *tierra caliente*, and therefore several large Andean cities have grown up at high altitudes; for example, Quito, the capital of Ecuador; La Paz, the capital of Bolivia; Caracas, the capital of Venezuela; and Mexico City, the capital of Mexico.

tierra templada (*Spanish*) The temperate areas of tropical Central and South America, which occur between 3,000 ft to 6,500 ft [1,000 m and 2,000 m] in the Andes near the Equator. Temperatures average 70°F [20°C] each month, and the total rainfall is about 60 in [1,500 mm]. Wheat and corn are grown as food crops; and coffee, cotton and a variety of fruits are grown as commercial crops. Much of the land is naturally forested, but when cleared it is suitable for agriculture. The original Indian inhabitants, the Incas, made terraces on many mountain slopes in the *templada* as well as in the *fria* zone.

till Glacial deposition; rocks, sand and fine-grained clay can all be classed as till. Some deposits are very sandy, while others consist mainly of clay, and some are very calcareous because the ice sheet passed over a large area of chalk. In the USA, the northern part of Long Island, New York, is till, and the southern part of the island is glacial outwash, separated from the till by a **terminal moraine**. Till is generally unstratified. ALSO CALLED **boulder clay**

tombolo A bar or **spit** which is joined to an island. Tombolos are not common.

topography The relief and configuration of a landscape, including its natural and man-made features. The hills, valleys, fields, forests, roads and settlements are part of the topography. A topographic map endeavours to show all such features, and needs to be quite large scale to show detailed information.

topological map A map in which lines are stretched or straightened for the sake of clarity, but without losing their essential geometric relationships. Well-known topological maps include the plans of the New York Subway, the London Underground and the Paris Metro. Sometimes called "rubber sheet geometry," the technique can be used for any route map and the end product will be simplified and clear.

topsoil The top layer or **A horizon** of the soil profile. It contains fine particles, **humus**, nutrients and the roots of plants. In areas of arable farming, it will be cultivated and turned over by plowing.

tor An outcrop of **granite** which stands above the general level of the landscape. It is an isolated mass of rock, showing vertical as well as horizontal lines of weakness where **weathering** has attacked the rock. It is a residual rock, the remnant of a layer of granite that has been largely worn away. It has survived either because it was harder than the surrounding rocks, or, more likely, because it was an area in which the **joints** were more widely spaced and therefore there was a slower rate of weathering. Around tors there are generally masses of weathered fragments, which are referred to as "clitter."

tornado l A small and intense atmospheric disturbance which travels across the countryside at 5 mph to 60 mph [10 km/h to 100 km/h] and contains winds of up to 200 mph [300 km/h]. Tornadoes are associated with intense heating in continental areas in late summer, when land masses are at their hottest. The center of a tornado is an area of low pressure which sucks up dust to give a blackish funnel rising to the sky. It is claimed that on many occasions cattle and humans have been sucked up by the funnel. Around this funnel of rising air are very strong winds which destroy crops and sometimes buildings. The low air pressure in the center of a tornado sometimes causes the walls of buildings to fall outward, because pressure is higher inside than out. Tornadoes are commonest in the interior of the USA, but also occur in India and many other countries. When tornadoes cross over water they become **waterspouts**. The average life of a tornado or a waterspout is 4 minutes. ALSO CALLED **twister** (*in North America*). **2** A violent storm at the beginning of the rainy season in West Africa. They bring strong winds and torrential rain, which is caused by mild air coming in from the sea and meeting the dry north-easterly air from the Sahara Desert.

tourism One of the major growth industries, tourism is important in many developed countries and is possibly vital to the economic development of poorer coun-

▼ **transect** A transect of the Atlantic Ocean showing the elevation of seafloor features. The vertical scale is greatly exaggerated to show features clearly – otherwise the transect would be nearly a straight line.

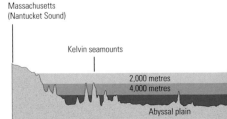

Massachusetts (Nantucket Sound)

Kelvin seamounts

2,000 metres
4,000 metres

Abyssal plain

tries. For example, southern Spain has developed rapidly since 1960 because of tourism, and Tunisia, the Gambia, Cuba and many less developed countries are now earning money from tourism. The attraction of sunshine is helping economic development in tropical and subtropical areas, and is an important factor in the high levels of tourism in Florida and California in the USA. Tourism has also grown into a major industry in snow-covered regions such as the Rockies and the Alps.

townscape The landscape of an urban area. It can be dominated by concrete and brick, but may also contain parkland. Cities in the USA have skyscrapers near their centers, but their suburbs may spread out over great areas, especially in the case of Los Angeles in California. European cities are less spread out, and contain large areas of suburban development. In New Zealand, where most residences are single story, urban areas spread out over fairly large distances.

trace element One of the elements present in the Earth's crust, which in very small quantities is essential for the normal development of plants and animals. For example, cobalt (for sheep and cattle), iodine (for humans), and zinc and copper (for plants).

trade Since spices were first brought to Europe from the Indies, international trade has grown steadily. Large quantities of wheat and cotton crossed the Atlantic from the USA to Europe, and manufactured goods were sent from Europe to all parts of the world. During the late 20th century, coffee was sent from Brazil to the USA and Europe, and oil from the Middle East was exported to Japan and Europe. Iron ore from Australia was sent to Japan. The wealthy industrial countries often buy raw materials from the less wealthy ones, but all countries are now dependent on trading links with other nations.

trade gap A higher level of imports over exports in a country's trade over a particular period of time.

trade wind A wind which blows from the tropical high-pressure zones (**horse latitudes**) to the equatorial low-pressure zones (**doldrums**). In the northern hemisphere, the air moving from the Tropic of Cancer toward the Equator is deflected to the right, making it a north-easterly wind. In the southern hemisphere, the air moving from the Tropic of Capricorn to the Equator is turned to the left to become a south-easterly wind. Trade winds are fairly persistent winds; they are all easterlies and tropical. They blow from deserts toward the tropical forest zone.

trading estate British name for an **industrial park**.

transect A line which cuts across an area. Transect lines are used in field studies; for example, to make a study of sand dunes, a transect line would be followed inland, across the dunes, and at right angles to the sea. Along the slope of the sand dunes

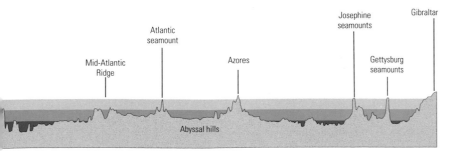

the changes of vegetation would be observed. The information gathered could then be mapped and a cross-section drawn along the transect line. A transect does not give as much detail as a complete study of an area, but it provides a sample without the necessity of too much time being spent on the task.

transhumance The seasonal transfer of livestock between mountain and lowland pasture. In the Alps, for example, cattle, and sometimes sheep, are taken up the hills to high-level pastures above the **tree line** during the summer months. Some members of the village community stay on the mountains in summer huts to look after the cattle, milk them and make butter or cheese. Meanwhile, down in the valley, other people cut the hay and look after any other crops. When the high-level pastures receive their first snowfalls in the fall, the cattle are moved back to the valley.

transition zone An area in an inner city where the older residential and industrial zones have decayed and are gradually being replaced by new developments.

transpiration The process by which plants give off and lose water vapor. It passes through stomata in the leaves, and goes out into the atmosphere. From the atmosphere the moisture may be recycled as rain, and then taken up into the plants again, via the roots. Transpiration contributes to the formation of rain. It has been discovered on the edges of the Sahara Desert in Algeria and also in the dry steppes of Russia, where millions of trees have been planted, that the rainfall is heavier than it used to be. Conversely, in parts of the Amazon basin in South America where forests have been totally removed, the annual rainfall has fallen from 80 in [2,000 mm] per annum to nearer 40 in [1,000 mm].

transportation network SEE **network**

transverse coast SEE **discordant coastline**

tree line **1** The line on a mountain above which trees do not grow. This is mainly because the temperatures are too low, but the quantity of rainfall or the lack of soil may also be influential. In the Andes, near the Equator, the tree line is just below 13,000 ft [4,000 m], and in the Alps it is about 10,000 ft [3,000 m]. The tree line is generally 3,000 ft to 6,500 ft [1,000 m to 2,000 m] below the **snow line** in temperate and tropical latitudes, though it will be much less in the higher latitudes. **2** A line of latitude beyond which trees do not grow, such as the northern edge of the **taiga**, where the forest gradually deteriorates and becomes **tundra**. **3** The lower edge of a forest below which the hillside is cultivated or covered with grass, as, for example, in many Mediterranean countries.

trellis-work drainage SEE **grid-iron drainage**

Triassic The first period of the Mesozoic era, lasting from 225 to 190 million years ago. Following a wave of extinctions at the close of the Permian period, many new kinds of animal developed. On land lived the first dinosaurs. Mammal-like reptiles were common and by the end of the Triassic period the first true mammals existed. In the seas lived the first ichthyosaurs, pacodonts and nothosaurs. The first frogs, turtles, crocodillians and lizards also appears. Plant life consisted mainly of primitive non-flowering plants, with ferns and conifers predominating.

tributary A subsidiary stream or river which flows into a larger river. Some rivers may have hundreds of tributaries, all of which collect the rain and ground water from a part of the river basin, but rivers which flow through deserts may have few tributaries; for example, the River Nile in North Africa. SEE ALSO **distributary**

tropical cyclone The type of violent **cyclone** characteristic of tropical areas in which wind speeds are likely to exceed 90 mph [150 km/h] and very heavy rain may fall. Tropical cyclones do not occur on the

Equator because there is too little air circulation; they occur mostly between 10° and 20° north or south of the Equator. They form regularly in the Bay of Bengal, causing flooding and devastation in Bangladesh. SEE ALSO **hurricane**, **typhoon**

tropical grassland Grassland found in tropical areas which do not have rainfall throughout the year, but only in the summer months. The total amount may be as much as 40 in [1,000 mm], but even so, forests will not grow because of the prolonged droughts of the winter months. A few **xerophytic** trees can survive, including several varieties of acacia, but the main natural vegetation is grass, which may grow to 6 ft [2 m] in height during the hot wet summer, but in winter it will go brown and shrivel up in temperatures of 60°F to 70°F [15°C to 20°C]. Pastoralists have to walk many miles in the dry winter to find pasture and water for their animals. Tropical grassland is found in Kenya, northern Nigeria, northern Australia, the Brazilian plateau and the Orinoco basin in South America. Most tropical grasslands are found between 5° and 20° north and south of the Equator, but in East Africa they are also found on the Equator, because the land is too high (and therefore cool) to support forest.

▼ **Triassic** During the Triassic period, the collision between the North America-Europe landmass and the Siberian plate pushed up the Ural Mountains.

tropical rain forest Luxuriant evergreen forest which grows in equatorial lands, mostly in the areas between 5° north and 5° south of the Equator. Rain falls most days of the year, and there is constant heat, so growth is rapid. Temperatures are from around 75°F to 80°F [25°C to 30°C] every month, and there is up to 100 in [2,500 mm] rainfall each year. The Amazon basin and Zaïre contain the largest expanses of tropical rain forest, but there are some smaller patches on the coasts of Colombia and northern Ecuador, northeast Brazil, the lowlands of Panama and Costa Rica, and the northern coastal region in Mozambique. Soils are generally poor and shallow in tropical forests. The nutrient for the trees comes from the dead leaves which fall throughout the year. The roots of the trees are quite shallow, and the larger trees have buttresses, which enable them to stand up straight and tall, often reaching a height of more than 100 ft [30 m]. Beneath the taller trees is a layer of smaller trees, which grow to 50 ft [15 m], and beneath them are the small trees of 15 ft to 30 ft [5 m to 10 m]. The trees form a dense canopy and very little sunlight can penetrate through to the ground, so that there is little or no undergrowth in tropical forests. The ground has millions of ants and other insects, and there is intense bacterial activity breaking down the fallen leaves, but there are few animals apart from the birds and monkeys found in the trees. There are many **lianas**,

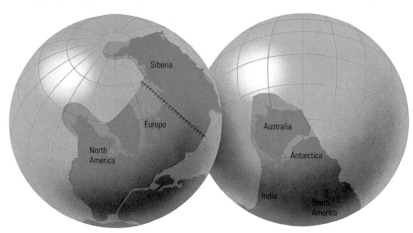

which are vinelike plants, and there are also epiphytes, which are parasitic plants found on tree trunks and branches. ALSO CALLED **equatorial rain forest**

Tropic of Cancer An imaginary line of latitude at 23.5° north of the Equator. It is the most northerly latitude reached by the overhead sun, on June 21, which is the **summer solstice** in the northern hemisphere. It is because the tilt of the Earth's axis is 23.5° that the sun reaches this latitude on its northward movement. The Tropic of Cancer passes through Mexico, Saudi Arabia, India and southern China.

Tropic of Capricorn An imaginary line of latitude at 23.5° south of the Equator. It is the most southerly latitude reached by the overhead sun, on December 22, the winter **solstice** in the southern hemisphere. The line of latitude passes through Australia, Paraguay, southern Brazil, and Botswana.

tropopause The boundary between the **troposphere** and **stratosphere**. The height varies, but it is about 10 miles [15 km] at the Equator and about 5 miles [8 km] at the poles. SEE ALSO **atmosphere**

tropophyte A plant that can live in both wet and dry conditions; for example, deciduous trees.

troposphere The lowest part of the **atmosphere,** below the **tropopause.** The troposphere contains the water vapor and hence all the weather phenomena of the atmosphere. In the troposphere there is generally a steady fall of temperature with an increase in height; this is called the **lapse rate**.

trough 1 A low-pressure area which is an elongation or extension of a **depression**. A trough of low pressure extends to the south side of depressions in Western Europe, and often contains a cold or occluded front. Troughs may bring rain because of the existence of the front, but they generally pass overhead within 12 hours. **2** A deep trench on the ocean floor; for example, the Puerto Rico (or Milwaukee) Deep in the Atlantic Ocean, at 30,249 ft [9,220 m], or the Mindanao Trench in the Pacific Ocean, at 34,439 ft [10,497 m]. **3** A deep valley, such as a glacial **U-shaped valley**.

truck farming SEE **horticulture**

truncated spur A spur which has been cut off by glacial erosion. Truncated spurs are found in glacial **U-shaped valleys** and the truncation leaves very steep valley walls. In some locations there are waterfalls, where streams flow over the edge of

TSETSE

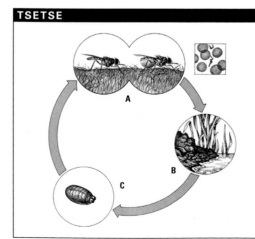

The life cycle of the tsetse fly *Glossina palpalis* involves three main stages. The abdomen of the adult fly fills with blood as it feeds through human skin (A) or bovine hide. Trypanosomes, which cause sleeping sickness, are often transmitted by an infected fly's saliva into the human or bovine blood at this stage. The female tsetse fly produces one egg at a time, which hatches inside her body. The larva is deposited on damp, shaded soil (B). It becomes a pupa (C), which develops into an adult fly.

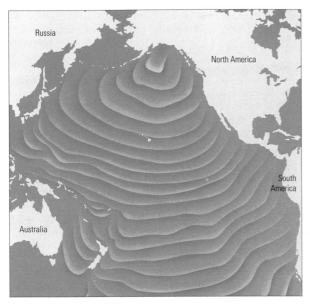

◀ **tsunami** When a submarine earthquake causes a sudden shift in the seafloor along a fault line, a tsunami is produced. The upheaval creates a bulge that causes a series of waves. The maps shows the hourly position of a tsunami that originated just south of Alaska.

the valley. At the foot of truncated spurs there are likely to be accumulations of **scree**. Truncated spurs can be seen in such places as Lauterbrunnen in Switzerland and Yosemite in California, USA.

tsetse Any of several blood-sucking flies in Africa, which transmit diseases to cattle and humans. To humans the tsetse passes on sleeping sickness, which produces lethargy and an inability to do any work; to animals it transmits the disease nagana. The tsetse occurs widely in tropical Africa, but needs woodlands for breeding purposes. Consequently, the open grassy plains are better areas for cattle rearing than clearings in the forests.

tsunami Ocean waves caused by an earthquake beneath the sea or a volcanic eruption. The waves travel at high speed across the ocean, often causing extensive damage when they hit land. SEE ALSO **earthquake**

tuff A rock which consists of consolidated ash and dust thrown out by a volcanic eruption. It is a soft and quite **porous** rock.

tumulus A mound built at the side of an ancient burial ground. The mound may be 30 ft to 60 ft [10 m to 20 m] in height and 60 ft to 150 ft [20 m to 50 m] in length. Tumuli date from pre-Roman times. ALSO CALLED **barrow**

tundra The subarctic area found to the north of **taiga** regions in Canada, Alaska in the USA, Norway, Sweden, Finland, and Russia. Summer days are quite long and often sunny, but average temperatures for July do not rise above 50°F [10°C]. For six to nine months the average temperature is below freezing point, and there is a thick layer of permafrost, of which only the top few inches thaw out in the summer. Where melting occurs, the land is wet and marshy, an ideal breeding ground for mosquitoes. The wet tundra soils are **gleys**, and waterlogging is characteristic. Generally, trees cannot grow in the tundra because of the unfavorable conditions, but there are some dwarf willow and birch in sheltered hollows. Mosses and lichens are common, and there are many small flowering plants in the summer. Reindeer (called "caribou" in North America), arctic fox, arctic hare and many geese, ducks and waders are present in the summer, but most move south in the winter. There is no real tundra in the southern hemisphere, though something similar is found in South Georgia and other Antarctic islands. High mountainous areas in the Himalayas, Andes, Norway and elsewhere also have areas which are similar to the Arctic tundra.

turbulence Irregular movements of the air in the lowest layers of the **atmosphere**. Turbulent activity causes mixing of the air, and often causes condensation as the rising air cools. Turbulence can cause showers of rain, as well as bumpy conditions for aircraft.

twilight The period of faint light before the sun has risen above the horizon in the morning and after the sun has descended below the horizon in the evening. It lasts only a few minutes near the Equator, but can last for an hour or more in the Arctic and Antarctic.

twilight zone An urban area where decline and decay have taken place, and there has been no redevelopment. It is often near the **central business district**, where old housing is in need of replacement.

twister A North American name for a **tornado**.

typhoon A tropical **cyclone** which affects the coasts of the Philippines, Hong Kong and China, generally in late summer or early autumn. Typhoons form over the China Sea as a result of excessive heating, and travel westward over the land, causing widespread destruction; they weaken once they move inland.

U

ubac The shady side of a valley, normally the north-facing side. It receives sunshine during the summer months only, and is likely to be forested or covered with grassland. SEE ALSO **adret**

Ullman, E.R., and Harris, C.D. Two American geographers who first described the multiple nuclei model of urban areas. They assumed that large cities have more than one central point or nucleus, and that growth spreads outward from separate nuclei to merge into one large urban area. They believed that urban expansion would swallow up nearby small towns, which would then create mini **central business districts**. Ullman and Harris improved on the **Burgess model** and Hoyt's **sector model**, but added a certain complexity.

ultrabasic An igneous rock containing very little **quartz** or **feldspar**. Ultrabasic rocks are usually **plutonic** in origin.

underdeveloped (*of a country*) Having little or no industrial development. The term is most frequently applied to **Third World** countries, where most people are still involved in agriculture of a subsistence type. A low average **gross national product** is possibly an indication of underdevelopment.

underground stream A stream which runs below the surface of the ground over all or part of the course. In some **limestone** areas, streams travel many miles underground, and such areas often contain **potholes**, underground channels and caves, in which **stalactites** and **stalagmites** form. Because the rivers change course from one **bedding plane** to another, there are many caves which no longer have rivers flowing through them.

undernourishment The condition that arises from an insufficient intake of food in order to maintain normal health. Millions of people in Latin America, Africa and Asia suffer from both undernourishment and **malnutrition**.

underpopulation Too few people in an area to fully develop the potential resources there. However, it may not be right, or necessary, to always stretch an area's resources to the limit.

unemployment Lack of employment; usually a state of being involuntarily out of work. The unemployment rate is calculated as the number of unemployed as a percentage of the total population of working age; a level of 3% or under is considered normal, allowing for people being unwell or changing jobs or areas.

United Nations An international organization founded in 1945 to promote peace, international cooperation and security. Other aims include the promotion of human rights and freedom, and to help solve economic, social, cultural and humanitarian problems. Originally there were 50 members; by 2001 this had risen to 189. It is the world's largest political organization and has six main bodies: the General Assembly, the Security Council, the Economic and Social Council, the Secretariat, the Trusteeship Council and the International Courts of Justice. There are then many specialized agencies and programmes which aim to help world development; for example, the United Nations Children's Fund (UNICEF), which provides basic healthcare and aid for children worldwide, and the Food and Agriculture Organization (FAO), which aims to raise living standards and nutrition levels in rural areas by improving food production and distribution.

urban Relating to towns or cities. An urban area can be defined as a built-up area, in terms of population density, or in terms of the total population of a settlement. Official definitions vary between countries. In Denmark, for example, any settlement with more than 2,000 inhabitants is classified as an urban area, and in India it is any settlement with more than 5,000 inhabitants. Any urban area should be multifunctional, and not just involved with agriculture, otherwise it would be called a rural settlement.

urban decay Deterioration and decay, especially of the older parts of a city. It is possible to restore or renovate these areas, and the older industrial towns of the USA, Canada, Europe and elsewhere have industrial buildings, canals and derelict housing, which are gradually being replaced. The areas suffering from greatest decay are in regions where factories have been closed and become derelict, and housing is substandard for the 21st century.

urban field The area around the city from which customers are drawn into the stores. A small city has a small urban field, extending for a few miles, but bigger towns have a field extending as far as 20 miles [30 km]. The distance will not be the same in all directions, as it will depend on the availability of roads or railroads, and also the proximity of other towns, which may draw customers in a different direction. The urban field of big cities such as New York, London or Paris can extend for hundreds of miles.

urban heat island A large city which is warmer than the surrounding rural areas as a result of the heat stored by the buildings and derived from the sun, and also as a result of central heating. An **isotherm** map will illustrate the phenomenon. In London, for example, the inner parts of the city may be 7°F [4°C] warmer than the surrounding rural areas, and outer parts will be 2°F to 5°F [1°C to 3°C] warmer. The higher temperatures mean that London has fewer frosts than most of southern and eastern England, and there is less snow, too.

urbanization 1 The growth and expansion of urban development and urban areas. **2** The movement of people into urban areas. Some examples of the proportion of populations classified as urban are: UK 92%, Australia 90%, USA 79%, Canada 78%, Brazil 71%, India 24%, Mali 23%, Kenya 17%, and Bangladesh 14%.

urban morphology The patterns of land use and the shapes of different zones which have evolved in many cities. It is possible to draw land-use zones for any urban area. There are three standard models of urban morphology, named after the researchers who described them: **Burgess model**, Hoyt's **sector model**, and **Harris and Ullman model**. However, every city is unique and will have its own distinctive features, even though there will be similarities with other cities.

urban renewal The replacement or renovation of declining older urban areas. Urban renewal is a slow and expensive process, but some towns, for example, Rotterdam in the Netherlands and Düsseldorf in Germany, which were badly damaged in World War II, have had opportunities for large-scale renewal schemes.

urban sprawl The spreading out of urban areas into surrounding rural areas as new houses are required for the growing population. The growth of the suburbs has been helped by the increasing number of cars, enabling people to live further away from their work in the downtown area. Before World War II, it was the buses and streetcars which enabled the suburbs to spread out further from the downtown area. The inter-war suburbs built in the 1920s and 1930s were generally along the main roads, sometimes as a form of **ribbon development**.

Urstromtäler (*German*) Wide trenches in northern Germany carved by meltwater from retreating ice sheets across the North European Plain in the last **Ice Age**. Today, these valleys are either filled with rivers or are useful in canal construction.

U-shaped valley A valley with a U-shaped cross-section. In their lower courses, rivers may build up wide **flood plains** which can be like a very flattened "U" in cross-section. Better U-shaped valleys are those which have been cut by glaciers, where ice erosion has flattened the floor and steepened the sides of the preexisting river valley; for example, in the Lake District in England, or the Alps. Glacial U-shaped valleys are often quite straight as the old river valley **spurs** have been truncated by the glacier. The floor will have been flattened, although in places it may have been over-deepened and gouged out to form a hollow. The hollow will be elongated, along the valley, and if filled with water will form a **finger lake**. The postglacial river flowing in a glaciated valley may be too small for the enlarged valley, and it will be called a **misfit**. Tributary streams may come rushing down the steep valley walls from **hanging valleys** high up on the hillsides. SEE ALSO **glacial erosion**

V

valley An elongated depression which is lower than the surrounding countryside, and is usually the product of river erosion. It may be no more than a few feet deep, or as much as 15,000 ft [5,000 m], as in the case of some Himalayan valleys. It may be just a few feet in width, or several miles; for example, the Mississippi or Amazon valleys. The cross-section of the valley may be a steep "V," a more flattened "V," or even a "U," and in many cases it will be asymmetrical. The shape will depend on the rock type and the amount of river erosion. The valley will have been eroded by a river, and the erosion may continue for millions of years. Some valleys have lost their river because it has been captured, or has gone underground, and they are called **dry valleys**. There are many examples in chalk and limestone regions. In mountainous areas, valleys are young and tend to be narrow and steep, often with rocky beds. In lower areas, the valleys widen out and become less steep. The **meanders** of the more mature rivers cause **lateral erosion**, which widens the valley. In lowland areas, where the rivers and landscape are in their old age, the valley has a flat floor, is very wide and has gently sloping sides.

valclusian A spring which emerges from underground in a limestone area. It takes its name from Fontaine de Vaucluse, a large and powerful spring in southern France.

veering A meteorological term denoting the clockwise change of direction of a wind, such as from southerly to westerly. SEE ALSO **backing**

vegas (*Spanish*) Irrigated lowland areas in Spain used for agricultural purposes, often orchards, and usually producing only one crop a year. SEE ALSO **huerta**

vegetation The plant life of a particular area. It will depend on the rainfall and temperature, and may be affected by rock and soil type. Humans may remove the natural vegetation, and use the land for cultivated vegetation, such as grass or food crops.

vein A thin deposit of a mineral occupying a crack or crevice in the rock. After volcanic activity, heated liquids and gases spread out through cracks in many of the surrounding rocks. The fluids contain minerals, which solidify as they cool down. Some will be metallic ores and precious stones; the majority are likely to be less valuable minerals such as **quartz** or **feldspar**. Lead, silver, zinc, gold, barytes, fluorspar, and many other minerals can be formed in this way, and they can be quarried or mined from veins.

veld A large expanse of grassland, found in southern Africa especially. It is the result of the climatic conditions which are not wet enough to enable trees to grow. Much of Northern Transvaal and Eastern Transvaal in the Republic of South Africa is veld, although in places the grass has been plowed up by farmers in order to grow wheat and corn.

vernal equinox SEE **equinox**

vertical corrasion Corrasion on a rocky riverbed which works downward to produce **potholes**. Gradually, the potholes become enlarged until they merge with their neighbors. In this way, the bed of the stream will be lowered.

vertical exaggeration The exaggeration of the vertical scale in a **cross-section** in order to show undulations. Although the horizontal scale is generally the same as that of the map from which the section is being taken, the vertical scale is likely to be much bigger. For example, a horizontal scale of 1 inch per mile (or 1 cm per km) will often be used with a vertical scale of 1 inch per 500 feet (or 1 cm per 100 m), which will give a vertical exaggeration of about (or exactly) 10 times.

vertical interval The height difference between neighboring contours on a map.

vertical link In manufacturing, one of a chain of sequential processes. In industrial organization, it is best to have vertical links within the same factory. For example, in textile manufacturing the cloth is spun, then woven, and finally made into a finished item of clothing in the same factory. This is vertical organization.

viscous lava Acidic slow-flowing lava, which forms steep mountain sides. **Basic lava** is less viscous and can flow quite swiftly, forming gentle slopes. SEE ALSO **acid lava**

viticulture The cultivation of grapevines. Grapes are mostly grown for making wine, but in some places they are used for currants, raisins and sultanas; for example, in Greece, Turkey and Australia. As well as the traditional wine-producing countries of France, Italy and Germany, many newer areas, such as California in the USA, Australia, and Bulgaria are exporting their wines.

volcanic bomb SEE **pyroclastic rock**

volcanic neck SEE **volcanic plug**

volcanic plug A hard mass of solidified lava in the vent of a volcano. Over a period of millions of years, the volcano will have been eroded, but the hard plug will survive as a small isolated hill in the crater. ALSO CALLED **volcanic neck**. SEE ALSO **puy**

volcanic rock Any rock formed on the surface of the Earth as a result of **igneous** activity. Because they form on the surface, volcanic rocks cool quickly and consist of small crystals, invisible to the naked eye. They are mostly basic in chemical content and form **basalt**. If they are acidic they form rhyolite. If they cool very quickly, as, for example, in a submarine eruption, they form obsidian or volcanic glass.

volcano An opening in the Earth's crust, through which molten rock, ashes, steam, etc., are ejected. Some volcanoes erupt in an explosive way, throwing out rocks and ash; others are effusive, and lava flows out of the vent; and there are volcanoes which are both explosive and effusive. The vent, or crater, is usually at the top of the volcano, but occasionally volcanoes blow out of the side of the hill. Some volcanoes have built up large

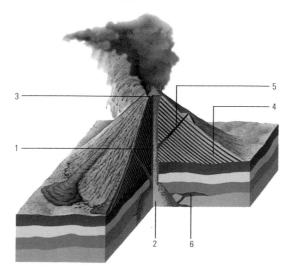

◀ **Volcanoes** are formed when molten lava (1) from a magma chamber (2) in the Earth's crust forces its way to the surface (3). The classic, cone-shaped volcano is formed of alternating layers of cooled lava and cinders (4) thrown out during an eruption. Side vents (5) can occur and when offshoots of lava are trapped below the surface, laccoliths (6) are formed.

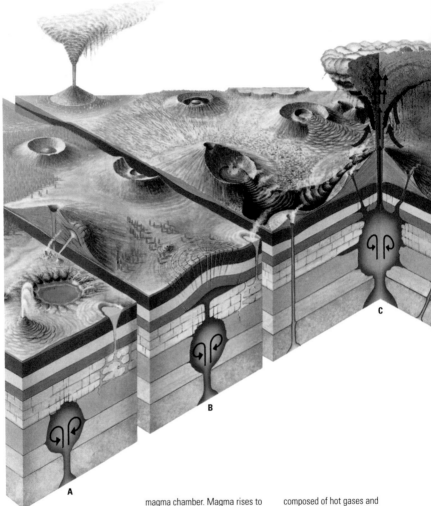

▲ **volcano** The forces that lie behind a volcano are born deep in the Earth. Mantle material upwells and becomes partly molten. The decrease in pressure as it rises causes it to melt even more and to melt the surrounding rock. The magma then ponds (A) about a kilometre below the Earth's surface and forms a reservoir or magma chamber. Magma rises to the surface (B) and erupts (C) when the pressure in the magma chamber exceeds the pressure of the surrounding rock. If the magma is viscous and the pressure drop is rapid, dissolved gases – mainly water vapour – explode out of solution. This blows the rock apart and sends the pyroclastic fragments high into the air, forming a massive eruption column composed of hot gases and incadescent pumice and ash. The particles heat up the surrounding air, causing convection currents that buoy them even higher – up to as much as 30mi (50km). When the column can no longer be supported by the surrounding air (D), it collapses to create incandescent pyroclastic flows that race outwards at velocities of more than 200 mph (100ms^{-1}).

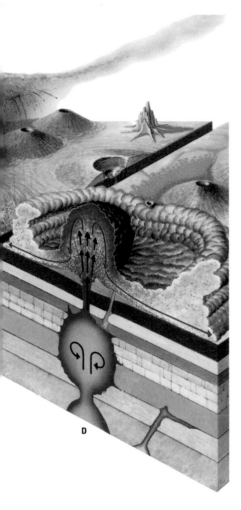

D

mountains as a result of many eruptions; for example, Mt. Etna, which has a circumference of over 75 miles [120 km]. Volcanoes may be extinct, which means they have not erupted for several decades; or active, which means they have erupted in recent years. There are over 600 active volcanoes in the world. The largest number can be found around the Pacific Ocean in an area known as the fiery ring of the Pacific; that is, in Japan, Indonesia, Philippines, New Zealand, Chile and the

western USA. The next most important area is the midworld belt running from the West Indies and through the Mediterranean. In the center of the Atlantic is the Mid-Atlantic Ridge, including Iceland, and there are a few minor areas; for example, in Antarctica and near the Great Rift Valley of East Africa.

There are different types of volcano according to the eruptions which caused them, and the shape of the mountain which has been formed:

1 Basic lava, Hawaiian type; for example, Mauna Loa. Free-flowing, runny lava has built up Mauna Loa into an enormous mountain. It is called a **shield volcano** because it is the shape of an upturned shield. Mauna Loa is over 30,000 ft [9,000 m] in height from its base, and has grown up from the bed of the sea to form an island. It is a higher mountain than Mt. Everest (at 29,029 ft [8,848 m]), but only 13,681 ft [4,170 m] is visible above sea level. Many basic eruptions come up through fissures, rather than through circular craters.

2 Acidic lava, **Peléan** type; named after Mt. Pelée in Martinique. The very viscous lava has formed a steep-sided volcano. In the big eruption of 1902, a spine of lava was left protruding from the crater, although it was worn away by erosion during the next two years. A large **nuée ardente** rolled down the mountainside and killed 40,000 people in the town of St Pierre at the foot of Mt. Pelée.

3 Strombolian volcanoes (named after Stromboli in the Lipari Islands, Sicily) are mildly explosive. Activity is frequent but usually quite mild.

4 Vulcanian volcanoes (named after Vulcano in the Lipari Islands, Sicily) erupt slightly more explosively, but less frequently than the Strombolian type. The lava is more viscous, and dark clouds and a great deal of steam are emitted.

5 Vesuvian volcanoes are even more explosive. They expel cinders and ash, as well as pouring out lava. Vesuvian volcanoes may be dormant for long periods, and then have a massive eruption. The lava is generally acidic.

Strombolian, Vulcanian and Vesuvian eruptions all give rise to composite volcanoes, as the cone-shaped mountains consist of layers of ash, cinders and lava.

6 The Krakatoan type is very violent, but may not give off any lava. Most volcanoes erupt basic lavas. Where the plates are moving together, the volcanoes tend to be more explosive and they erupt acidic lava, together with cinders and ash; for example, Japan, Philippines, Indonesia, and Mt. St Helens in the western USA.

von Thunen, J.H. A 19th-century German estate-owner, who lived near Rostock in eastern Germany. In 1826 he wrote a book called *Isolated State*, in which he outlined his ideas about the economics of land use. Using a farm or a village as a central place, he believed that the most intensive farming, dairying and market gardening would take place in a concentric zone near to the central place. Further away from the central place, there would be another concentric ring of less intensive types of farming, because it would take the farmer longer to walk out to the fields. In this area he would grow crops which required little attention. Beyond the second ring, there would be a third concentric ring of very extensive farming, where the land, crops and animals needed very little time spent on them. Von Thunen's ideas of **location theory** in agricultural land use were based on the assumption that the land was flat and featureless, and that transportation facilities would be equal in all directions. This is called an **isotropic** plain and does not really exist. He also assumed that there would be only one form of transportation and one market. However, his general ideas still have some relevance, on various scales and in various locations, all over the world. Around farms and villages in many **Third World** countries, the distinctive zones of decreasing intensity can be seen. In European farming, also, the most intensive areas are often in the center, such as near London, Paris or Brussels, while less intensive farming areas can be found further from the major towns.

V-shaped valley A valley with a V-shaped cross-section formed by river erosion. Both vertical and lateral **corrasion** take place.

vulcanicity The process connected with the formation of, and the movement below and through the Earth's crust, of magma and other associated materials.

W

wadi A narrow steep-sided valley found in deserts and semiarid regions. Wadis are dry valleys for much of the year. Rainfall is not frequent in desert areas, but, when it does rain, it falls in heavy showers. As much of the land is bare with no vegetation, **surface runoff** can be rapid. The water will carry sand and small rock fragments, and so a great deal of erosion takes place. **Vertical corrasion** can form narrow steep-sided valleys, which can increase to several hundred feet in depth. In humid lands, there is more frequent rainfall and rainwash to wear away the valley sides, making them more V-shaped in cross-section. Some wadis extend for long distances, and it is possible for a heavy storm in one location to cause flooding in a wadi which can rush for many miles, possibly into areas where there has been no rainfall. Such rapid and short-lived floods have been known to catch travelers unawares. Wadis are common in Egypt, Libya and many other desert areas. ALSO CALLED **arroyo**

warm front A leading edge of warm air. The warm air will be rising gradually over the colder air which has gone ahead. As the air rises it will cool and form cloud, probably **cumulus** or **cumulonimbus**. Eventually it may give rise to some warm rain, which will be frontal rainfall. A **depression** often contains a warm front, and it is the first front to arrive. It will be followed a few hours later by a **cold front**. In the northern hemisphere the warm front is generally located to the south of the center of the depression. SEE ALSO **occluded front**

warm sector A region of mild or warm air situated between the warm and cold fronts in a **depression**. It is generally cloudy with **stratus**, and may give some light rain or drizzle, but is often dry. It lasts for only a few hours before the **cold front** arrives to give heavier rainfall and lower temperatures. In most depressions, the cold front gradually catches up and merges with the **warm front** to form an **occluded front**; the warm sector will then have disappeared.

Warsaw Pact A defence and economic alliance formed in 1955 by Albania, Bulgaria, Czechoslovakia, East Germany, Hungary, Poland, Romania and the USSR as a response to the **North Atlantic Treaty Organization (NATO)**. Albania left in 1968, and the rest of the Warsaw Pact was wound up in 1991 after political reorganization and change in Eastern Europe and the USSR.

water cycle SEE **hydrological cycle**

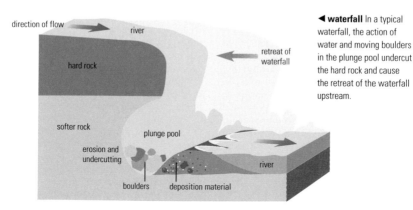

◄ **waterfall** In a typical waterfall, the action of water and moving boulders in the plunge pool undercut the hard rock and cause the retreat of the waterfall upstream.

direction of flow

river

retreat of waterfall

hard rock

softer rock

plunge pool

erosion and undercutting

boulders

deposition material

river

WATERSPOUT

A waterspout occurs when a tornado passes over water, sucking up a column of water up to 300ft (100m) tall. They usually occur in tropical regions, associated with stormy weather.

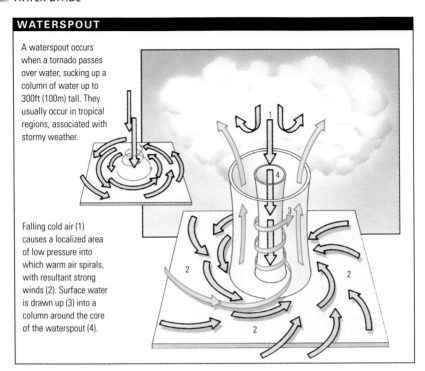

Falling cold air (1) causes a localized area of low pressure into which warm air spirals, with resultant strong winds (2). Surface water is drawn up (3) into a column around the core of the waterspout (4).

water divide SEE **watershed**

waterfall A steep or vertical descent of the water of a stream or river. Where there is a change of rock type on the bed of a stream, there are different rates of erosion. This creates an irregularity on the **long profile** of the stream and forms rapids or a waterfall. The softer rock is eroded more quickly, causing an abrupt change in the gradient. If there is a stratum of hard rock which is vertical or nearly so, only a small waterfall will be created. If the harder rock is horizontal, a much larger waterfall can form. The surrounding softer rocks may be worn away and may possibly undercut the hard rock. If this happens, there will be a collapse, though the waterfall will not disappear. It will just move upstream, and gradually a gorge will form downstream from the waterfall. A good example is the Niagara Falls on the US/Canadian border.

water gap A gap in a ridge cut by a stream or a river. A famous gap in the USA is the Delaware Water Gap in eastern Pennsylvania, where the River Delaware has cut through the sandstone of the Kittatinny Mountains, forming a gorge about 2 miles [3 km] long.

water parting SEE **watershed**

water power The power produced by the movement of a body of water. For example, a stream can turn a water wheel, which can drive a machine. Water power was used in the early cotton and woollen industries of Lancashire and Yorkshire in Britain, and New England in the USA. It has been used for centuries in mills which grind wheat into flour. In many parts of the world, water from rivers is used to drive turbines and generate electricity in hydroelectric power stations. **Hydroelectric power** is the most important type of electricity in some countries; for example, Norway, Sweden and Switzerland. There are large schemes in countries all over the world; for example, Iguaçu in Brazil and near Bratsk in Russia.

watershed The high ground which forms the boundary or dividing line between two river basins. On one side of the boundary line the water will drain in one direction; on the other side of the boundary line it will drain in the opposite direction. The watershed is normally a ridge or a piece of ground that is higher than the surrounding areas. It is often a very irregular line. ALSO CALLED **water divide**, **water parting**

waterspout A funnel-shaped mass of water, which is similar in formation to a **tornado**. It occurs over water when a heated patch of air rises, and whirls around in a circular or corkscrew fashion. Waterspouts are mostly seen in tropical areas, especially in late summer. When waterspouts go on to land they become tornadoes, and when tornadoes go over lakes or the sea, they become waterspouts. Waterspouts are generally less than 100 ft [30 m] in height, and they last

for half an hour or so. Often a few occur on the same day and they may be accompanied by strong gusts of wind. Large boats can sail through them undamaged, but smaller sailing vessels would be knocked over.

water table The level below which the rock is saturated. After rain has fallen, the water gradually trickles downward through the pores in the rocks. The level at which all pores are full is the water table. The height of the water table gradually changes, moving up or down depending on the recent rainfall. After heavy rain the water table will rise, and in spells of wet weather it could reach the surface, causing puddles or lakes to form. On average the water table lies a few feet below the surface. The line of a water table generally follows the shape of the land surface, although it is normally neither as steep nor as high. Water located below the water table is called **ground water**.

wave In water (usually oceans), the swell or forward movement caused by the fric-

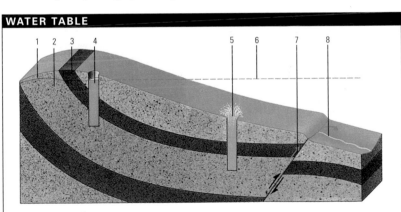

WATER TABLE

Artesian springs and wells are found where groundwater is under pressure. The water-table (1) in the confined aquifer (2) lies near the top of the dipping layers. A well (4) drilled through the top impervious layer (3) is not an artesian well because the head of the hydrostatic pressure (6) is not sufficient to force water to the surface. In such wells, the water must be pumped or drawn to the surface. The top of an artesian well (5) lies below the level of the head of hydrostatic pressure and so water gushes to the surface. Artesian springs (8) may occur along joints or faults (7), where the head of hydrostatic pressure is sufficient to force the water up along the fault. Areas with artesian wells are called artesian basins. In the London and Paris artesian basins, the water has been so heavily tapped that the water level has dropped below the level of the well heads.

tion of the wind on the water surface. The strength and size of the waves will be dependent on the wind speed and the distance of open water across which the wind blows (the **fetch**). Waves cause coastal erosion when they reach the shore; as the wave nears the shore it breaks, because of friction with the seabed, and carries debris up the beach as swash. It then flows back to the sea as **backwash**.

wave-cut platform An outcrop of level or gently sloping rocks on the shore, which have been scraped and smoothed by **abrasion**. The sea carries sand and pebbles, which scrape away at any exposed rock surfaces. A wave-cut platform is often created at the foot of cliffs and may gradually increase in size as the cliffs retreat inland because of erosion. ALSO CALLED **abrasion platform**

wave refraction The bending of a wave as it approaches the shore, due to a variation in the depth of the water. If the waves are coming in diagonally to the coastline, the part of the wave in shallower water will slow down, and the crest line of the wave will begin to curve. Waves also bend near headlands because of the same effect caused by shallowing water.

weather The meteorological conditions which are being experienced over a short time period. When the weather is recorded over a few days and then averaged out, it can be called **climate**. The weather is the result of the state of the **atmosphere**, and the pressure, wind, temperature, humidity and rainfall caused by the atmospheric conditions. Near the Equator the daily weather is much the same, day after day, and so the weather is almost the same as the climate. In deserts and near the poles, the daily weather may be similar for a few days or even weeks. In many other areas there may be frequent and rapid changes of weather.

weathering The decay and disintegration of rocks caused by the effects of weather and atmosphere. Weathering is

one of the major processes involved in denudation and causes the breakdown of rocks *in situ*. No movement or transport of the weathered material is involved; otherwise, the process would be classified as **erosion**. Weathering may be mechanical, chemical or biological. Biological weathering can be the result of roots opening up cracks in the rocks, and worms, rabbits, bacteria, etc. can also contribute to rock disintegration. Chemical weathering may take several forms. Rainwater contains carbon dioxide from the atmosphere and can cause solution of salts in the rocks, or carbonation in chalk and limestone. Oxidation is the result of rainwater and oxygen from the air leading to the disintegration of ferrous minerals. Hydration is another type of chemical weathering, in which a mineral combines with water. Mechanical weathering (also called physical weathering) is mainly the result of heating and cooling. The heat of the sun causes rocks to expand, and when they cool at night they contract. The expansion and contraction of minerals causes rocks to crack. If the temperatures fall below freezing point, the process is more effective. Any water in the rocks will freeze and expand by 9% of its volume. Alternate freezing and thawing will lead to the disintegration of any rock. SEE ALSO **exfoliation**

Weber, Alfred Weber described a standard theory of industrial location. He believed that the optimum location for an industry would be where transportation costs were lowest. Collecting raw materials and distributing the finished product would be the vital consideration. In the case of the steel industry, the best location would be somewhere between the sources of iron, coal and limestone. The availability of a labor supply and the location of markets would also be important. The ideas in Weber's locational model are still economically sound, though conditions have changed appreciably since he wrote in 1919. In the case of the steel industry, most new works are so large that no one place can provide all the raw materials. Because

▶ **weathering** The breakdown of rock in place is called weathering. It occurs in two main ways: physical (A and C) and chemical (B). They usually occur in combination. At the surface, plant roots and animals such as worms break down rock, turning it into soil (A). In chemical weathering (B), soluble rocks such as limestone (1) are dissolved by groundwater, which is a very mild solution of carbonic acid. Acid rain caused by sulphate pollution (2) also attacks the rock. The water can create cave systems deep below the surface. Both heat and cold can cause physical weathering (C). When temperatures drop below freezing, freeze-thaw weathering can split even the hardest rocks, such as granite (4). Water that settles in cracks and joints during the day expands as it

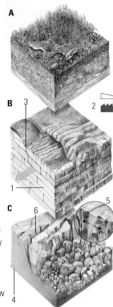

freezes at night (5). The expansion cleaves the rock along the naturally occurring joints (6). In deserts, rock expands and contracts due to the extremes of cooling and heating from day to night, resulting in layers of rock splitting off.

of this, coastal locations are generally used so that large quantities of materials can be transported by sea, the cheapest form of bulk transportation. In many other industries transportation costs may be quite low, especially if few materials are required and the finished product is quite small. SEE ALSO **location theory**

wedge An area of high pressure extending from an **anticyclone**, similar to – but narrower than – a ridge of high pressure.

welfare geography The part of **human geography** which studies social inequalities; it looks at how the welfare of people

differs spatially and at the factors which contribute to the quality of human life (such as crime and poverty).

West, the A general and shorthand term used to describe North America and Western Europe.

westerlies The major planetary winds in **temperate** latitudes. There are two main types of large-scale planetary winds, which blow out from the tropical high-pressure regions; they are the westerlies and the **trade winds**. The westerlies blow toward the poles from the **horse latitudes**, but, because of the rotation of the Earth, become south-westerlies in the northern hemisphere (for example, near Britain) and north-westerlies in the southern hemisphere (for example, near Tasmania or southern Chile). While the trades are quite persistent, the westerlies are very variable. Most of the weather in temperate latitudes comes from the west, but it includes **depressions** and **anticyclones**, in which the winds will be blowing in different directions. The effects of the continents and oceans interferes with the flow of westerly weather in the northern hemisphere, but there is less interference in the southern hemisphere, which is mainly ocean. In the northern hemisphere, westerly weather affects the USA, Canada, Britain, Iceland and Norway for much of the year, and in winter, when the world's winds all move southward, westerlies can affect the Mediterranean lands, too, bringing winter rainfall.

wet bulb thermometer SEE **hygrometer**

wet day A day on which measurable rain has been recorded.

wet site The site of a settlement where water is available in an otherwise dry area, such as a well in a desert, or a spring in a limestone or chalk area.

wheat A **cereal** grass of the genus *Triticum*, which is widely cultivated in temperate areas. It grows best in areas with more than 120 frost-free days, where temperatures are 63°F to 68°F [17°C to 20°C] in the last two months of the growing season, with a dry spell for harvesting. Total rainfall requirements are 15 in to 30 in [400 mm to 700 mm], depending on the season of rainfall and the amount of evaporation. Major growing areas include the **prairies** and the **steppes**. It is the major cereal in temperate latitudes.

whirlpool A circular movement of water in a river or the sea, caused by the meeting of two currents or by the shape of the rocks on the bed of the river or sea.

whirlwind A whirling column of air caused by a small rising current of hot air, which may pick up dust and cause a dust storm.

white-collar worker A nonmanual worker who works in a clean environment, such as an office. It is a term also used to describe professional people. SEE ALSO **blue-collar worker**

white ice Ice on or near the surface which still contains some air. At greater depths all the air is removed by compression, and the ice will look blue. As the ice changes from white to blue, there is melting followed by freezing, which is called **regelation**.

wind The movement of air from one place to another. Air moves from high pressure to low pressure in an attempt to balance out the pressure differences. As the wind travels, it is diverted to its right in the northern hemisphere, or to its left in the southern hemisphere. If the **isobars** are close together, the pressure gradient will be steep, and the wind will be strong. More widely spaced isobars mean there is a low-pressure gradient, and the wind will be gentle. Winds are named by the direction from which they are blowing, so a southwesterly wind actually blows toward the northeast. Winds are recorded by anemometers

and can be measured in knots, miles per hour, or meters per second. The **Beaufort scale** is a traditional method of measuring wind speed.

windbreak A line of trees planted to provide shelter from strong or persistent winds. Many houses in the Rhône valley in France are on the south side of clumps of trees, which protect them from the **mistral**. On the **prairies** and the **pampas**, rows of trees are frequently planted around the farmhouses. Some windbreaks are planted to protect soil and prevent erosion.

windchill The perceived cooling effect of wind in low-temperature conditions. The air temperature feels much colder if there is a wind blowing. The physiological temperature is the way in which the human body feels the temperature of the air, and this is strongly influenced by the windspeed and direction. Temperatures below freezing feel very cold if the windspeed is 5 mph [10 km/h] or more, and if the wind reaches 25 mph [40 km/h], temperatures of 40°F [5°C] will also feel very cold.

wind energy Power generated from controlling the wind, usually using windmills. It is one form of alternative energy that could be used to generate electricity, but it is fairly unreliable given the inconsistency of the wind.

wind gap A gap, **col**, or small notch in a hilltop, through which the wind can whistle. The gap would have been carved by water erosion, possibly from a glacial meltwater stream. A wind gap may be the result of **river capture**; it will not have been formed by wind.

wind rose A diagram used for showing wind direction at a weather station. The principle points of the compass are shown, and the length of line on the rose indicates the number of days the wind has blown from a particular direction. The number of calm days is generally written in the center of the rose.

WIND ENERGY

A wind generator converts the energy of the wind into electricity. By far the most common type is a wind turbine and a generator placed on top of a narrow concrete tower. Three-bladed, variable pitch designs are the most efficient. Variable pitch means the attitude of the blades (1) can be changed. By altering the pitch (2) of the blades, they can generate at maximum efficiency in varying wind conditions. The whole rotor assembly rotates (3) into the wind. The blades turn a prop shaft (4), which links to a generator (5) through gearing (6). Large groups of such wind generators are known as "wind farms", and the largest of these have thousands of linked turbines that can produce the same amount of power as a fossil fuel power station.

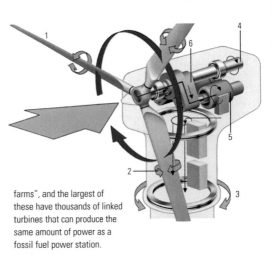

windward A term used to denote the side of a hill or building which faces into the wind. The opposite is "leeward."

winterbourne A stream which flows in winter but dries up in summer when the water table is lower. Such streams are found mainly in chalk areas and the name is generally used only in southern England.

winter solstice In the northern hemispher, December 22; but June 21 in the southern hemisphere. SEE ALSO **solstice**

wold A chalk hill in Lincolnshire or Yorkshire in England.

World Trade Organization On January 1, 1995, this multilateral trade organization replaced the **General Agreement on Tariffs and Trade (GATT)**. There were 135 members at the end of 1999.

World Wide Fund for Nature (WWF) The world's largest environmental charity, which was founded in 1961 as the World Wildlife Fund to raise money to help in the conservation of many different species. Its aims have now been widened to include the conservation of natural resources and environments, as well as wildlife.

wrench fault SEE **tear fault**

xerophyte Any plant which is adapted to survive in arid or semiarid conditions. Some xerophytes have long roots to reach underground supplies of water; others have small or thick leaves, or no leaves at all. Some have fleshy stems in which to store water, and many have a thick bark or a waxy layer to reduce transpiration. **Xerophytic** plants such as cacti are found in deserts, and thorny bushes, such as sage brush, are found in arid and semiarid regions. Palm trees survive in many places, and tough tussock grasses occur widely. Many flowers grow in deserts after a rainstorm, and the plants go through a rapid life cycle. When the plants die, seeds are left to lie dormant until the next rain falls.

xerophytic Able to withstand drought conditions.

Y

yardang A narrow steep-sided ridge found in various arid areas; for example, Turkestan in Central Asia. Yardangs often occur in groups with several ridges running parallel to each other and oriented in the direction of the prevailing wind. They are the result of sand **corrasion**. Wind blowing between the yardings removes any depositional material which is accumulating.

young mountain A mountain that has been recently formed. The young mountains include the Rockies, the Alps and the highest mountains in the world, the Himalayas. Fold mountains formed by alpine **earth movements** have undergone only 40 or 50 million years' erosion and, on a geological time scale, this makes them a young landscape feature. Although the alpine folding created several large mountain ranges, there were a few smaller hills formed around the edges of the main ranges.

youthful 1 (*of a river valley*) A river in its upper or mountain course; the valley is narrow and V-shaped in cross-section and active vertical erosion will be taking place. **2** (*of a landscape*) A landscape still covered with mountains; for example, the Rockies, the Himalayas or the Alps. As the landscape is eroded, the mountains become lower and the valleys wider, and the landscape gradually becomes mature or middle-aged.

yurt SEE **gher**

Z

zero population growth The point at which there is no increase in population. Zero population growth occurs when the **birth rate** and **death rate** are approximately the same. This point has been reached by Sweden, France and Britain. When a country has an optimum population, it will hope to achieve zero population growth. Many **Third World** countries have a population increase of 3% per annum, and the world average is about 1.3%. Accordingly, there is much to be done in order to achieve zero population growth throughout the world. Many countries still receive immigrants, and this will maintain an increase in population even when the birth rate and death rate are roughly similar.

zeugen A pedestal rock with a tabular-shaped profile caused by wind erosion. Variations in resistance to erosion are responsible for the irregular profile, but there is normally greatest erosion just above ground level, where wind and sand activity are at their maximum. Examples can be seen in most deserts, but especially in Central Asia. SEE ALSO **pedestal rock**

zonal soil A soil found in large areas having a similar climate and vegetation. For example, climatic conditions which are ideal for grass in the **prairies** and **steppes** have contributed to the great accumulation of **humus** which created the **chernozem** soils. Zonal soils fall into one of two main groups: **pedalfers** and **pedocals**.